EG Practice
Renewable Energy in the Countryside

Second edition

Peter Prag

Estates Gazette Group, 1 Procter Street
London WC1V 6EU
A division of Reed Business Information

Commissioning Editor: Alison Richards
Typesetting: Isabel Micallef – in Meta and Palatino
Cover and interior design: Deborah Ford, EG
Printed and bound by: Hobbs
Cover images: © Digital Vision

Contents

Acknowledgements, iv

Preface, v

1 **Introduction**, 1
Sources of energy, 1
International treaties, 2
UK policies, 2
The Renewables Obligation, 3
Energy from land, 4

2 **Saving the planet**, 5
Climate Change Programme, 5
The emergence of wind power, 6
New trends in farming and forestry, 8
Environmental measures, 11

3 **Harnessing the wind**, 13
Commercial developments, 13
Planning – initial considerations, 18
Planning – environmental issues, 21
Single turbines, 23
Action points for farmers and landowners, 24

4 **Biomass**, 25
Climate change, 26
Commercial production opportunities, 29
Cultivation, 31
Biomass from forestry, 33
Small-scale production, 35
Action points for farmers and landowners, 35
Action points for woodland owners, 35

5 **Biofuels**, 37
The UK market, 38
Biodiesel, 40
Bioethanol, 40
Biobutanol, 41
Bio-oil, 41
Biomethane, 41
Action points, 42

Appendix 1: Trade organisations, government agencies and other sources of information, 43

Appendix 2: Glossary, 49

Index, 53

Acknowledgements

First edition

The subject matter is wide ranging and largely new and still unrecorded. I am particularly grateful therefore to have been able to consult a number of people who have an active interest in the matter, including the following: John Ainslie of npower, Richard Barker of Wind Prospect, Peter Clery of BABFO, Rosalind Clifford of Smiths Gore, Campbell Dunford, Peter Edwards of Windelectric, Gary Freedman of Ecotricity, David Harbottle of Faccombe Estate, Raymond Henderson of Bidwells, Alison Hill of BWEA, Malcolm Shepherd of Wessex Biofuels, and Tim Shuldham of Coppice Resources Ltd.

Second edition

Many further developments have occurred since the first edition was published at the beginning of 2005 and to incorporate these into the present text I have drawn extensively on websites, a number of which are listed in the directory. I was also grateful to speak to individuals engaged in these various fields, including: Gill Alker of TV Energy, Christopher Maltin of Organic Power Ltd, Hugo Marriage of H Dolton & Son Ltd, and Josh Pollock of Fisher German.

Preface

The Government is committed to developing the production of renewable energy: first in generating electricity and now also in biofuels. Where this is derived from onshore facilities, it will involve the use of land either for the installation of turbines or for growing crops for power stations or for processing into transport fuel. Meanwhile, agricultural incomes have fallen and the Common Agricultural Policy has been reviewed so that more farmers are having to look at alternative uses of their land. Getting involved in renewable energy is one such alternative, but it is a new and relatively untried area that comprises a wide range of issues that need to be properly assessed.

Government policies and financial inducements have created artificial and possibly short-term markets in a field that depends on long-term commitments. They have also produced pressures on developers and planning authorities and given rise to imbalances between the various potential systems. Farmers and landowners have every reason to become involved but before doing so they need to be well informed about the various opportunities and difficulties that they are likely to encounter.

This publication outlines the current policies and markets for renewable energy within Britain and considers the main factors that are involved in putting them into practice.

Introduction

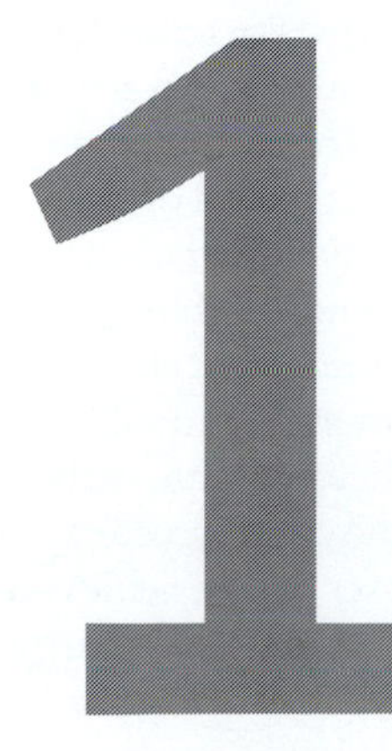

Sources of energy

The expression "renewable energy" sounds like magic, as if the energy that you consume is renewed and available to be used again. In reality, it is rather more prosaic and implies merely that the material from which the energy has been generated is renewable. This may be either because it is derived from a naturally recurring phenomenon, such as wind or sun, or from something that can be replaced, as in the case when timber or arable crops are replanted after harvesting.

This is not a new concept, but it has recently taken on a new perspective and impetus. Wind and water have been used for centuries as a source of power for milling flour and then for industrial processing, as well as for irrigation. For centuries too, coppice was harvested as a fuel, for both domestic purposes and processing. In time, coal and then oil and gas replaced these traditional resources, coupled more recently with the harnessing of nuclear power. Latterly, however, this "balance of power" has been disrupted by a series of events, some of which were sudden and dramatic while others may have been more gradual but are nonetheless of lasting influence. Many of these occurred far from the UK, demonstrating that domestic energy policy is determined often by global factors.

In the Middle East, in October 1973, Israel took swift and successful action against its Arab neighbours in what was to become known as the Yom Kippur War. In revenge, the Arab nations stopped supplying oil to Israel's allies and the price of crude rose fourfold.

This resulted in a widening of the search for alternative sources of oil and gas in what would previously have been untenable locations, such as Alaska and the North Sea. Also, it enabled an expansion of nuclear power plants and encouraged the development of new alternative forms of energy, notably solar panels, although the technology for this was still of only limited potential under the British climate.

Then, in the Ukraine, during the night of 26 April 1986 a malfunction occurred in a power station near the town of Chernobyl, about 80 miles north of Kiev. The nuclear fallout from this "melt down" spread over thousands of miles and lasted many years, with residues being found in sheep flocks in North Wales and Cumbria for more than a decade afterwards.

It seems to have had another, longer-term outcome in that there is resistance within the UK to the concept of producing electricity from nuclear power. No more such power stations are being planned within the UK and the existing ones are going to be gradually decommissioned. These policies may be based upon sincere beliefs, but they seem to ignore the possibility that Britain may not be able to sustain a viable electricity supply in the future without recourse to some nuclear production. The British Government has only recently taken a fresh view of this, commissioning an *Energy Review* that was published in 2006 and which concluded that nuclear power would have to be an essential part of the country's future energy requirements. Even if this currently controversial proposal were to be adopted, the time needed for the development of new nuclear power stations is such that there is likely to be a national shortfall in electricity before they can be completed. Meanwhile, there will be an increasing focus on energy saving and on all other forms of supply, including from renewable resources.

International treaties

In Brussels, in 1992, the European Commissioner for Agriculture secured agreement for a series of reforms of the Common Agricultural Policy aimed at tackling the overproduction of foodstuffs. The MacSharry Plan, as it was known, introduced the concept of setaside, whereby farmers were subsidised for leaving land fallow or alternatively using it for non-agricultural crops, which included, for example, those that could be processed into industrial oils.

This new regime incorporated also a Mid Term Review, which was instigated in 2003 and has furthered the original policy by "decoupling" support from agricultural production entirely. Within this too are a number of grant schemes aimed at diversification and rural regeneration some of which give rise to opportunities for using farmland for the production of biofuels and electricity.

Meanwhile, in Japan, in December 1997, the story had moved even further afield to the hitherto little-known city of Kyoto. An international summit of nations was assembled in order to address the issue of climate change, or "global warming", and to tackle the causal problem of carbon dioxide emissions. Carbon dioxide was seen as one of the main "greenhouse gases" that have been damaging the ozone layer and thereby contributing to global warming. It is produced extensively in the exhaust of all vehicles and industrial processes that use oil or gas, which have furthermore been recognised in themselves as being finite resources that need to be conserved and therefore consumed more frugally.

A protocol was drawn up setting targets for the proportion of energy that signatories would aim to produce from renewable energy as opposed to fossil fuels, such as coal, gas and oil. Some major countries, such as the US and Russia, refused to implement these proposals but, in 2000, the European Union (EU) did so through the introduction of the Climate Change Programme.

As a member country, the UK has adhered to this with a series of measures aimed at reducing the level of carbon dioxide emissions by 20% below the 1990 levels by 2010. It is these measures that bring the story firmly back home to roost.

UK policies

They are in fact crucial to the concept of renewable energy within Britain, since without them there would not have been the political will to create the regulatory and economic conditions upon which the industry now depends. Their practical significance within a global context, however, seems to be tiny, to the point of being irrelevant. In the base year of 1990, Britain produced 569 million tonnes of carbon dioxide, or about 2.8% of the world's total output. The latest figures show this to be 2.3%, as global emissions rose to around 24 billion tonnes.

If the UK were to succeed in meeting the 20% reduction required by Kyoto by 2010, it will have made a difference of only 0.5% to the world's output of greenhouse gas. To achieve this, this will have involved increased energy and production costs in the UK, while some other nations continue without such restraints. This may seem unreasonable in terms of the national economy and industry, but for the rural sector, which is to be the main provider of carbon-saving energy, it should present a unique and positive opportunity.

The imbalance between Britain and the rest of the world is currently exacerbated by those major countries that remain outside the Kyoto protocol, notably the US, China and India, who, between them, account for about 10 billion tonnes of carbon dioxide per annum, or around 40% of the global total. This failure to comply stems from doubts primarily within the US over the science relating to global warming and to a lack of political and economic will to impose constraints upon energy use.

Russia was eventually able to subscribe to the terms of the protocol thanks to the financial advantage that could be gained from the high level of emissions that prevailed in the base year of 1990 at the end of the Soviet era. However, in the relatively newer economies of India and China it would still be unfeasible to incorporate the costly cleaner production methods required. With the rapid development of industry in both China and India, the imbalance in global carbon emissions can only worsen over the coming years.

The reduction in greenhouse gases can be achieved by a number of means, including particularly energy saving in transport, construction and industry, for which there are also various forms of Government support. The main focus within Britain has, however, been on the development of energy systems. There is a wide diversity of such systems under consideration, ranging from photovoltaic cells to hydrogen fuels, although many of these are still in a relatively early stage of development and are unlikely to be in commercial operation before 2010. Nuclear power, which was due to be decommissioned in Britain, is now being reconsidered within the Government's *Energy Review*. This covers a range of issues of energy production and saving and sets out proposals for achieving emissions reductions by 2020. The more immediate considerations within the UK, however, are focused partly on transport fuels but essentially on electricity generation.

The Government's present commitment, or "aspiration" as it has latterly been called, is that 10% of the country's electricity should be produced from renewable sources by 2010, with the hope that this might rise to 20% by 2020. There are regional variations to this, set by the devolved administrations, as for example in Scotland where the target figures are 18% by 2010 and 40% by 2020. This reflects the different climatic and geographic conditions as well as a different political situation. Current estimates indicate, however, that the proportion of electricity produced by renewable means, including hydroelectric as well as wind power, lies between less than 1% to a maximum theoretical capacity of about 4% (the differential in these figures is due to the variability of wind power, as mentioned in chapter 3).

It would seem then that much has yet to be done. A fundamental problem is that electricity is more expensive when produced from any means other than conventional systems using fossil fuels or even nuclear reaction. In a free economy, the additional cost of developing and using alternative production methods therefore has to be balanced either by subsidy or by regulation or by a mix of the two.

The Renewables Obligation

The mechanism by which this might be achieved within the UK is the Renewables Obligation (RO), which was introduced in April 2002 (and also the associated Renewables (Scotland) Obligation) to replace an earlier instrument known as the Non-Fossil Fuels Obligation (NFFO). The workings of this arrangement are explained in some detail in a later section, but the basis upon which it operates is that electricity suppliers are now required to show that a certain proportion of their production derives from a renewable resource.

The evidence for this is in the form of certificates that are either granted to the suppliers by the generating companies from whom they have sourced their electricity or may otherwise have been purchased from other suppliers who have a surplus. Failing that, they would need to pay a "buyout" price for any shortfall to the industry regulator, Ofgem. That "levy" is then ultimately repaid to those suppliers who have certificates, the amount being in direct proportion to their holding for the year.

This creates a mix of regulation and subsidy and, at March 2003, the required proportion was set at an initial rate of 3% of total production, increasing in stages to a target of 15.4% in 2015. The figure for 2006/07 is 6.7%, but actual production is currently likely to fall short of this by about 85%, or at best by 40%, so that there is a growing pressure within the industry to acquire the necessary certificates. There is pressure too on commercial users of electricity who are required to pay a Climate Change Levy on any power that they consume that is not certified as coming from a renewable source, in which case they would be granted an exemption. Ultimately, the cost of this arrangement is paid for by the consumer and this has been politically feasible due to the relatively low cost of electricity at the time. Since then, however, the price has risen not only due to the effect of the RO, but also following a steep increase in the cost of oil and gas around the world.

The main new source of electricity is from wind, primarily from turbines built on land but now also from installations out to sea. Large-scale hydroelectric power is also excluded from the RO on account of its cost and to its potential due to location and other constraints, and the use of biomass in power stations is another qualifying process, although it has as yet to attain viability. These various production methods are referred to in later sections, but for the present it is worth noting that each of the present available systems depends on one essential ingredient, namely land.

The Climate Change Programme is not exclusively about electricity, and efforts are being made to produce vehicle fuels from materials that are cleaner and more readily replaceable than mineral oils. These can be produced from arable crops and then blended with conventional fuels so as to operate safely within existing vehicles while reducing the level of carbon emissions and also reducing the amount of oil and gas extracted from mineral reserves.

The commercial development of such fuels has been hampered in the UK by the fact that they cannot initially be produced at prices that would be competitive with mineral oils. The Government sought to rectify this in 2004 by granting a concession in the level of excise duty payable on biofuels, but the measure was then postponed until 2005 and the proposed allowance of 20p per litre was reckoned by the industry to be too small to bridge the price gap. Since then, the Government has introduced an RO for transport fuel in 2005, similar to that being used for electricity, to be implemented in 2007 with a target of having 5% produced from renewable sources. This will require the market place to part fund the production of biofuels, as mentioned further in chapter 5. If fuels are then to be produced from growing crops, it will depend again on one essential ingredient: land.

Energy from land

Landowners and farmers are therefore crucial players within this venture and without their involvement the programme would surely founder, or be sent entirely offshore. Offshore both in terms of the location of wind turbines and in the supply of biomass which could be imported as oilseed or woodchips. Land used for the erection of wind turbines is acquired through private negotiation and not by some form of compulsory arrangement. Biomass is sourced from timber residues or from coppice grown under contract on agricultural land, and energy crops for the production of biofuels are also harvested from arable production. Opportunities exist not only for sales to energy suppliers, but also for individual on-farm units that exploit those same natural resources as a "home-grown" alternative to buying heat and light from a national network.

All this arises at a time when agriculture within the EU is undergoing a major restructuring and change of direction, with an increasing emphasis on environmental criteria and diversification. It would seem that farmland is no longer required predominately for food production and, similarly, that home-grown timber is now rarely viable in its traditional form. Meanwhile, political agencies throughout Europe, and, in particular, the British Government, are struggling to implement an environmental programme that depends largely upon a new utilisation of land.

Might this be a new utopia for farmers and landowners or a potential minefield of difficulties and disappointments? This book aims to identify the reality of what may be involved.

The commitment to developing alternative sources of energy comes at a time when both farming and forestry are able to diversify into new forms of production that could include fuel crops. In the UK, Government policy has focused more on generating electricity, due largely to the fact that it is less costly in terms of Treasury funding, although there are significant limitations in the efficiency of wind turbines and in meeting rural planning requirements.

Meanwhile, the opportunities for growing energy crops that had previously been thwarted through lack of funding is now beginning to become a commercial reality.

Saving the planet

Climate Change Programme

Agricultural diversification

Farmers and landowners are well accustomed to the concept of introducing some non-agricultural element to their properties and businesses. Indeed, the latest Government figures show that over half of farmers within the UK now have an alternative source of income and diversification schemes are becoming an increasingly common sight across the country. In almost all such cases, the development will have been undertaken essentially for financial reasons, either to supplement failing farm incomes or to exploit new opportunities. There are only few occasions when part of a farm may be given over to another use without some monetary motive, although this does arise in certain situations, such as Countryside Stewardship or creating a visitor centre for school children. Diversification schemes do often create a facility that benefits a wider public, as in the case of farm shops or pay-as-you-go bridleways, but, essentially, they will be founded upon a commercial purpose.

The economics of agriculture and land ownership in the UK have been so depleted in recent years that there has been little scope for benevolent acts of providing public facilities or environmental enhancement. Farmers are not heartless about such matters and are generally by their very nature and experience highly aware of environmental and other countryside issues. The majority, however, have not felt able to invest their own time and money into acts of common good unless they were to receive some financial return, which may either be in the form of a commercial income or as grants or perhaps a mix of the two. Therefore, any policies or proposals regarding the production of renewable energy from farmland or forests will inevitably need to provide a sufficient financial return if farmers or landowners are to consider introducing it alongside, or in lieu of, existing output. There may be a powerful case for increasing our consumption of renewable energy from systems that are largely dependent on the use of land, but the owners and occupiers of that land are unlikely to be in a position to respond to these global concerns without some form of financial involvement.

The cost of energy

The development of alternative sources of energy will also need to be assured of a financial return, especially while conventional forms of power remain relatively cheap and highly competitive. Therefore, the implementation of a Climate Change Programme depends on more than just political will, in that there are substantial costs to be borne as well; costs of development and costs also to the consumers.

The global balance sheet of renewable energy is a complex one that extends beyond the actual cost of developing new processes. Conventional producers, such as the oil and gas industry, can be expected to resist the potential threat to their existing markets and consumers will not be easily convinced of the need to accept more expensive tariffs for fuel and power. Rising energy costs have a direct impact on inflation and international competitiveness and so on the economy as a whole. It is not easy then for governments to introduce measures to encourage the development of alternative forms of energy if these will be loading additional costs onto consumers, particularly at a time when the price of oil and gas has been increasing.

On the other hand, if steps are not taken to reduce the consumption of fossil fuels, there may be different costs incurred through the consequences of global warming and the likelihood of shortages in vital energy stocks. The Kyoto meeting of 1997 achieved a remarkable consensus among nations to the effect that measures needed to be taken to reduce the consumption of fossil fuels, but in the end no full agreement was reached and some major participants, such as the US and Russia, did subsequently withdraw from the protocol. For them, and some others such as China and India, the debit side of the balance sheet appeared to have been too daunting. However, Russia has latterly acceded to the agreement, which means that it could be ratified as a treaty as it did then include a prescribed minimum number of signatories.

Carbon emissions

Nonetheless, the fact that some major industrial nations remain unwilling to adopt the Climate Change Programme makes it harder for others, such as those in the EU, to do so. Harder in terms of international competition and harder too in terms of whether the measures implemented by the remaining signatories will have a sufficient effect on climate when they appear to be ignored in the US, which has just 4% of the world's population is said to produce 25% of total global emissions. There are increasing doubts too in some quarters as to whether emissions from fossil fuels are in fact a cause of global warming and even whether such climate change is really taking place.

Meanwhile, there could be some economic advantage for those nations that do adopt climate change measures in that countries and companies may be able to trade emissions credits among each other under the Emissions Trading Scheme. Organisations that have invested in energy-saving technology could gain a financial return from it if they operate within the minimum targets and so have spare "capacity" to sell on elsewhere. (The required reduction in emissions is to be measured against 1990 levels, when Russian industry was still operating under the Soviet system and was highly pollutant. It may be then that they will by now be well placed to gain some valuable credits and this was part of the rationale in making what would otherwise have been a costly political decision to subscribe to the Kyoto Agreement after all.)

Carbon trading has already reached the countryside, as in the case of Country Grampian Pork. In 2005, this company, which has over 40 pig units spread across Britain, acquired from British Airways the right to emissions for 2000 tonnes of carbon. Pigs, in common with most other livestock, produce a high concentration of greenhouse gases largely in the form of methane. Aeroplane engines of course also emit exhausts of carbon dioxide. Both companies had signed up to a voluntary scheme under which they were each allocated a maximum level of emissions. If they exceeded that level, they would have to pay for a licence to cover that excess; if they used less than their entitlement, they could sell the surplus on an open market. British Airways found that they were subsequently operating beneath their allocated level, due to a number of factors including the introduction of more efficient engines. Grampian needed to expand and were able to buy the surplus from British Airways as a more economic alternative to paying for an additional licence. Almost as if "pigs could fly"!

There is scope too in the rural context for developers of wind farms to utilise the fact that their installations are carbon neutral and then sell credits in the market for raise development capital. Furthermore, a growing number of companies are committing themselves to offset as far as possible the carbon emissions that arise from their businesses and this could also be directed into the countryside. The emissions may arise from, for example, the essential workings of an industrial process or the air travel or road transport employed by that business. One potential offset has been found in planting trees, which absorb carbon, as is explained in chapter 4, but which would depend on finding an appropriate area of land. Another might be providing finance for construction of a wind farm.

The emergence of wind power

The UK Government is currently committed to implementing a Climate Change Programme as set out in the European Directive. The principal target is currently that 10% of the national electricity should be generated from renewable resources by 2010. In Scotland, a target figure has been set at as much as 18%. There are policies also for encouraging the development of biofuels and research is in hand on other energy sources, such as from hydrogen cells, solar panels and waves. The priority, however, has been for electricity generation, primarily from wind turbines, but also with the possibility of fuelling power stations with biomass.

Focusing initially on electricity generation has had a number of attractions. The working technology has already been established, both for wind turbines, largely in Scandinavia, Germany and the US, and for biomass conversion in France and elsewhere. It should therefore be possible to introduce these systems into the national network more rapidly and at a lesser cost than in the case of developing new techniques such as wave power. There are other advantages too in that the British Isles is relatively well positioned for wind and that the turbines are highly visible and therefore a clear demonstration that the Government's policies are being effected.

These factors also create drawbacks; being so visible the turbines make what is often considered an unacceptable impact on the landscape and the British climate is not so wild as to allow the turbines to operate for more than around 30% of the time. The intermittent nature of the supply of electricity generated from wind raises two further problems when trying to balance that supply with demand. The first is of a mechanical nature in receiving the input from wind into the grid and the second concerns the manner in which electricity is traded across the country.

Financing renewable electricity

The New Electricity Trading Arrangements (NETA) were introduced in 2001, with the aim of trying to even out supply and demand by means of financial incentives and penalties in the contracts for generating and supplying power. Raising or lowering output from a conventional power station is reasonably feasible, especially to meet the normal seasonal variations, but controlling the supply generated by wind turbines is more difficult, particularly as there is still no means of storing electricity on this scale.

At present, generators of renewable electricity are entitled to operate outside NETA and to deal direct with suppliers. This may reduce the financial penalties that might otherwise be incurred, but the physical limitations still remain. There are drawbacks too with using biomass in that the process is more efficient at producing heat than electricity, but creating heating plants requires greater investment than seeking initially to adapt existing power stations.

The financial incentive to produce electricity from means other than fossil fuels derives from the Renewables Obligation (RO), introduced in the Utilities Act 2000. In essence, this is the antithesis of a subsidy in that it is a Government instrument whereby suppliers of electricity are compelled to deliver a specified percentage of power from renewable resources. Any electricity that has been produced from an approved source is authenticated by the issue of a Renewables Obligation Certificate (ROC). While the amount of power being generated by these means is limited, and is indeed less than the percentage demanded under the Obligation, suppliers have to acquire the necessary Certificates, which thereby command a premium price. This enables the generators who produce electricity from renewable sources to offer attractive terms to those third parties upon whom that process depends, such as the owners of sites on which wind turbines are erected. The equation then depends upon the proportion of power that has to be matched by ROCs and the time period over which this policy is being implemented. The premium price paid by the suppliers has meanwhile been charged to the end consumer, rather than funded by Government.

When first implemented in 2002 to replace the failing NFFO, the RO was set initially at a rate of 3% of total supply and programmed to rise annually until 2010 when it would be 10.4%. Production at that time was at only about 1.2% of UK consumption, so Certificates were bound to be in demand and attract premium prices. A difficulty arose, however, with the term over which the Obligation had been implemented, as there was no guarantee that the system of Certificates would remain in place after 2010. It was possible therefore, even if unlikely, that the premium prices would not be sustained beyond that date. Seven years could be insufficient time within which to plan and develop a wind farm and to recoup one's investment, particularly if it had to be sited in a relatively inefficient location due to planning constraints as mentioned below.

By 2004, the percentage of supply requiring Certificates had risen to 4.3% and the Obligation was extended to 2015, by which time the rate would be increased to 15.4%. This ambitious target for renewable energy production was coupled with new initiatives to facilitate the development of offshore wind turbines, as mentioned in chapter 3.

If the amount of electricity produced by renewable means continues to fall short of the number of ROCs required by the suppliers, then the value of those Certificates is likely to rise. The ensuing higher prices could then mean that sites that were previously thought to be uneconomic, due to their sheltered location and low average wind speeds, could yet become viable. The price of ROCs, which have been trading at up to about 5p per unit depending on the length of contracts, has been kept reasonably stable by means of notional certificates. These have been made available by the Government in order to overcome what would otherwise be a shortfall in electricity from renewable sources that would qualify for ROCs and have been issued at a capped price of currently a little over 3p per unit. The proceeds of these sales are then redistributed to those suppliers that have achieved the full percentage target, by way of a bonus. This mechanism helps to make up the shortfall between the amount of renewable energy being produced and the target level required under the RO. The system relies upon being able to pass the cost of ROCs to the consumer. At its introduction in 2002, wholesale electricity prices were at their lowest in real terms for over 30 years. Since then, they have increased by around 75% and continue to be on a sharply rising trend. It remains to be seen whether such an arrangement, which loads additional charges on to consumers, can be sustained at a time of high basic costs.

In 2006, the Government gave a note of warning within its *Energy Review* that it would be seeking to consult over the level of financial support arising from the ROC system, suggesting that it be "banded" as of around 2009 according to the degree of need. This could result in payments being reduced for onshore developments so as to provide more help for offshore projects instead. At around the same time, the National Grid also gave warning that it might be seeking greater financial contributions from wind farm developers towards the cost of providing a connection from remote wind generators to the point of demand. Meanwhile, another cost has risen through the imposition of business rates on commercial turbines. These new developments could affect the viability of some future onshore developments and are also considered further in chapter 3. Meanwhile, another cost has arisen with the imposition of Business Rates on commercial turbines.

There is a further aspect to the market operation of renewable energy in that the Government has introduced a Climate Change Levy (CCL) that is imposed on all commercial and institutional users of electricity, initially at the rate of 43p per unit. Those users who can demonstrate that they have acquired electricity from a renewable source are granted exemption from paying this levy. This comes in the form of a Levy Exemption Certificate (LEC), which is issued by Ofgem, the official industry body, who also administers the ROCs. Suppliers of electricity who have had to acquire ROCs and to raise the overall price of their electricity to cover the cost of those Certificates are able then to secure a reduction in total price for those of their customers who have purchased renewable electricity and benefited from the exemption. This is effectively a partial subsidy against the cost of ROCs, with the distinction that the need for ROCs is determined by statutory regulation and is unavoidable whereas the obtaining of LECs would be a voluntary measure.

Uncertainties over how these market factors are likely to operate may lead to consumers continuing to use conventional electricity and to pay the levy accordingly. Those that do choose to acquire power from renewable sources, for reasons of economics or more usually policy, are doing so only on paper, in the form of certificates. It cannot be supplied physically direct from some wind farm (which is as well since that would then be only intermittent and virtually impossible to use).

Planning

In the present context, it becomes clear that there are serious constraints upon the workings of the existing policy. The time factor is one particular issue affecting the economic viability of wind farm developments, but another concern is that of efficiency.

If an investment is to show an adequate financial return, the turbines must be able to produce a certain minimum amount of electricity. For that they need to be sited in windy locations, which are necessarily on exposed sites; but exposed not only to the wind but also to view. Not only are such sites highly visible, but many are also in relatively rare landscapes, such as mountain or moor that have been designated for their special qualities as in the case of National Parks or Areas of Outstanding Natural Beauty. This has implications on the planning process but the most immediate consequence is that gaining planning permission becomes more difficult and more costly, not only in money but also in time. That time can be crucial as it erodes the limited period during which the development can be assured of viability under the RO.

Planning is often the single most important issue in harnessing wind power within the UK and is considered in the chapter 3. It happens also to have been raised in the *Energy Review* of 2006 as an area that needs to be streamlined if projects are not to be unduly delayed or deterred. While that may have been aimed more at major construction schemes such as new nuclear power stations, it could, if implemented, have a positive impact on the further development of in-shore wind farms.

New trends in farming and forestry

Agriculture

Farming in Britain and throughout Europe is experiencing a major period of transition following the Mid Term Review of the Common Agricultural Policy (CAP), which was introduced in 2003. This has heralded a radical change in the manner in which financial support is provided to farmers, with a switch in emphasis away from payments based on the area being cultivated or the volume of crops being produced.

Subsidies are now being determined more according to the overall extent of the farm, rather than what is being grown on it, and upon various environmental and other conditions being fulfilled. Irrespective of cropping or production, farmers now become eligible for a flat rate support payment provided that they have been able to comply with these conditions and have maintained their land to the prescribed standards.

In England, the change in system is being phased over a period of seven years, whereas in Wales and Scotland the new arrangements were adopted in full at the outset in 2005 and Northern Ireland is using a hybrid of the two. In any of these cases it is likely that the level of payments will be reduced by degrees in the future under the principle of "modulation". This is a mechanism that enables governments to cut back the rate of payments and switch the proceeds to other forms of rural development. Farm income will otherwise be dependent entirely on the market value of the produce being grown on the holding and on other earnings, such as from diversification. This is widely expected to coincide with an era of more free world trade, which could lead to a lowering of prices for basic foodstuffs.

Many farming systems that had been profitable prior to these political changes will need to be reviewed and restructured if they are to continue to be viable. Therefore, there will be considerable pressure upon farmers to consider diversifying out of conventional crops.

Industrial crops

Many farmers may leave agriculture altogether, by selling up or letting their land to neighbours. Others are increasingly looking to gain an alternative income from outside farming whether from the property itself, such as letting holiday accommodation, or from working elsewhere. Leasing sites for wind turbines would fit well within the former category, but it is only available to those relatively few landowners whose properties are situated in an appropriate location.

There will be others, particularly in arable areas, who could switch from food production to growing materials for biofuels. The basic crops can mostly be grown using the same arable techniques and equipment as before and so not require any new skills or investment, at a time when the latter could be difficult to finance. Growing biomass, such as willow, on the other hand, does need a greater degree of commitment in equipment and skills, as well as time since it takes longer to establish than an ordinary annual crop. Some forms of biomass, which are detailed in chapter 4, could be derived from materials that are already being produced on the farm, notably straw and poultry litter.

Having the skill and equipment to switch into industrial crops will not be sufficient in themselves to counter declining returns from conventional farming. The new production system must also be capable of making an adequate financial return. This will depend on a number of factors, not least the level of prices being offered by processors during the initial development stages. Those prices will also be dependent on many other factors, such as development costs and the general state of the fuel markets as well as the availability of grant aid and tax incentives. In the case of one group of products, the price achieved by growers will also crucially be determined by one particular factor, namely that of the cost of transport.

Biomass

Biomass, in the form of either harvested coppice or forestry residues, is a relatively bulky commodity and current industry practice suggests that coppice, for example, is unlikely to be viable unless it can be grown within about 25 miles of a power plant. If the price of electricity and other forms of power were to rise substantially, then no doubt the economics of converting biomass would improve accordingly and be able to cover the higher cost of transporting material over greater distances.

This is not expected in the immediate future and the wider use of land for growing biomass will rely upon more power stations being developed within an increasing number of locations. At present, that appears to be hampered by the nature of government policy and regulation covering the development of such installations, with the result that little progress is currently being made in this field. This is mentioned in more detail in chapter 4.

Biofuels

Similar constraints affect the further development of biofuels, such as diesel and ethanol. As mentioned above, farmers would be readily able to use their land to grow conventional arable crops for processing into transport fuels rather than food. There would appear to be every incentive to do so in a financial environment in which agricultural returns are set to decline. All they would need is to be able to sell their produce at a better, or at least equivalent, price than that which they receive for foodstuffs.

The price that the processors can offer to the growers is determined by the market price for petrol and diesel and the cost of converting crops into those fuels. In the initial development stages, the cost of conversion is relatively high against the level at which established oil companies are able to trade. The retail price of fuels is determined, in some degree, by the international oil markets, but also largely in the UK by domestic tax. In 2004, the British Government introduced some concessions in duty for biofuels, which under current market conditions have enabled the industry to begin to embrace the crucial start-up costs. As explained in chapter 5, this is now starting to offer a viable alternative use for arable land.

Setaside

Some non-food crops, such as those for fuel, have been cultivated in Britain for industrial purposes, but this has been generally on land that is in "setaside". The CAP continues to require all but the smaller arable farmers to keep a proportion of their land (currently 8%) uncropped or setaside.

Under the new system, there is now no separate payment for setaside, but it forms part of the conditions of compliance for receiving CAP support. In most cases, the land would be left fallow for a year and managed simply for basic weed control. However, it is permissible to grow non-food crops on such land and still receive the overall payments, and thereby create a double income. By this means, it has been feasible to grow, for example, oilseed rape for processing into industrial oils at a price that would not on its own compete with that being paid for the food varieties. The fact that this is restricted to only 8% of a farmer's arable area limits its commercial potential, although such income can be boosted through the receipt of separate grants that have been introduced specifically to encourage the cultivation of energy crops, as itemised in a later section.

Conservation

With farming in Britain being under such pressure and as the returns from agriculture decline due primarily to the CAP and international trade agreements, a growing number of farmers and landowners have sought to enhance their incomes from other means, often putting part of their properties to a new and non-agricultural use. This trend towards diversification has wrought changes in the landscape as barns are converted to offices or workshops and pastures become pony paddocks.

This has also engendered a degree of acceptance among both the public and the planning authorities that the countryside is changing and that traditional farming scenes may have to include commercial and industrial elements. This may make it easier now to countenance, for example, the construction of wind turbines in open country, which might have been unthinkable just a few years previously. The increasing spread of mobile phone masts could be an indication of current attitudes about erecting what are effectively industrial structures in the countryside.

On the other hand, while conventional farm incomes have declined, conservation and the natural environment have both gained importance, in terms of public policy and grant aid as well as in the public consciousness. This has led to an ethic of protectionism and resistance to development in the landscape, driven by pressure groups and local communities, which can act as a significant counter-balance to the view that the countryside must modernise and diversify. New developments, such as road construction, have been thwarted or delayed by the perceived need to preserve a particular species of flora or fauna that can be as small and seemingly uninteresting as a type of lichen or snail. Wind turbines bring their own conflict between the desire on the one hand to protect the landscape and on the other to improve the environment by enabling the use of renewable energy. The position of the landowner in all this can be difficult and is considered further in chapter 3.

Forestry

Income from forestry in Britain has also been in decline, as timber values have fallen by around two-thirds during the past 10 years, according to Forestry Commission figures. This is due to a number of factors, principally perhaps the relatively strong level of the pound against other currencies, which has enabled timber to be imported at a lower price than that which has to be charged for home-grown produce. Indeed, each year the UK tends to import around 85% of the country's timber requirements, despite the fact that there are approximately 2.8 million hectares of woodland throughout Britain, with a potential for producing about 14 million tonnes of timber annually.

In recent years, market prices have been below the level at which timber might show a viable return and many growers have delayed felling or thinning their plantations as a result. This is particularly true of the poorer-quality species, such as sitka spruce, which was planted so extensively across upland areas.

There is little reason to expect any significant recovery in domestic timber prices within the foreseeable future and the situation is likely to continue to be problematic. Furthermore, the age profile of most of Britain's commercial woods is such that the volume of maturing timber is expected to increase considerably during the next 20 years, from around 12 million to 20 million tonnes per annum. This is due to the extent of new planting that took place under the more encouraging economic and political conditions that prevailed around 30 and 40 years ago.

New markets

There is every reason for foresters to look for new initiatives that might enable them to dispose of what is likely to be an increasing supply of largely poorer-quality timber. At present, such timber is converted into pulp or pallets or may, together with sawmill residues, be chipped for use in the manufacture of fibre boards. Those products will probably become oversupplied as more plantations reach maturity and the timber for those markets would then be difficult to sell. This will comprise not only trees that were to be felled having reached maturity, but also the thinnings that would normally have been taken during the years preceding maturity. It is already often no longer viable to carry out such thinnings, when sale prices are below the cost of doing so. The forester then misses out on what would have been an interim income from the plantation and the quality of the final crop will be reduced, so it will become even harder to sell the timber at full maturity.

These crops could be suitable for processing in power stations and there would seem to be a strong rationale for developing systems for converting this material into energy. Precedents for this do exist elsewhere in Europe, notably in Scandinavia where in Finland around 20% of energy is derived from wood fuels.

The technology for processing wood fuels is already available but, as yet, it has had only very limited application within Britain. One feature of it has become more widespread throughout this country in recent years and that is the increasing use of chipping machines within woods. Whereas previously after trees had been felled the waste material, such as branches, would have been burnt on site or left to decay on the ground, it is now more often ground up into chippings or mulch. These chips can, if required, be gathered up and be used as a potential fuel.

Environmental measures

In addition to the global issues considered at the beginning of this chapter, there are certain specific environmental factors that may impact on the development of renewable energy within the countryside. One overriding issue is the fact that Government policy has been focused largely upon generating electricity rather than other forms of power, such as heating and transport fuels. Biomass can be an efficient provider of heat but is less effective when used solely to generate electricity.

In other countries, new power stations have been developed to produce Combined Heat and Power (CHP) for local communities and public or commercial installations. The localised nature of CHP may require new concepts and new investment in the context of power plants, infrastructure and public services, but it does thereby avoid the problem of transporting power over large distances. Electricity is often generated at some remove from its main consumers, especially in the case wind power, and linking it to towns and industrial areas involves expense, loss of efficiency and a further intrusion into the environment in the form of pylons and grid lines.

In the UK, the use of biomass is still emerging only slowly after an almost disastrous start, as also mentioned in chapter 4. However, it could be better suited to rural communities than to towns, where the construction of CHP plants would involve a relatively higher value site and where questions over emissions and the transporting of bulk fuel might be more difficult to resolve. These issues are likely to be easier to accommodate when constructing and servicing a plant in country towns or villages, many of which do not have mains gas and would therefore gain an added benefit of a communal system of heating.

Co-firing

Coal-fired power stations produce particularly heavy emissions of carbon dioxide and other gases. One means of reducing this is to adapt them to co-firing whereby the coal is mixed with a less damaging material, notably wood in the form of chipped pellets and sawdust. Most coal-fired stations can absorb a mix of up to about 5% wood fuel without having to make any significant modifications. Even this relatively small percentage would have a worthwhile impact on emissions, as biomass produces only around 7% of the amount of carbon dioxide that would come from coal when generating the same unit of heat.

The rate at which co-firing is being taken up is still restricted, due partly to a requirement that if the generators are to qualify for ROCs, the wood fuel must be stored and blended with coal on site rather than bought in ready mixed from an outside supplier. Power stations have to be able then to invest either in storage for the wood fuel or in blending equipment if they are to benefit from the value of the ROCs. Since wood fuel, and especially sawdust, absorbs water if left in the open, which then reduces its efficiency when combusted, the storage facilities have to be covered against the weather and are thereby are more costly. That expenditure would then have to be recouped over a relatively short period in that the arrangements for enabling co-firing to be eligible for ROCs will close in 2016. This time-scale is linked to a planned phasing out of coal-fired power stations. Were it to be financially more feasible meanwhile to adopt the process of co-firing, then it might enable the production of biomass to be developed for such time as it would be required for use in CHP plants.

Alternative sources

There are other possibilities for producing renewable energy, such as from the sun by solar panels and photovoltaics or from tidal power, but most of these are still in early stages of development and are unlikely to involve any significant use of land. Hydroelectric power is already in use, but is generally confined to the more mountainous parts of the country and does not usually concern landowners other than for smaller private schemes. Furthermore, larger hydroelectric schemes can encounter adverse environmental issues arising from the flooding of wide areas of land. This potential difficulty is reflected in the fact that these schemes will no longer be eligible for subsidy in the future, despite working with a totally renewable resource, such as water, and being almost entirely carbon neutral. There are systems also for utilising ground source heat, whereby the constant temperature of subsoil or rock is pumped into a heat exchanger. The technology is, however, still limited to small-scale installations

Municipal waste can be another source of energy that utilises land, whether in the form of methane being drawn from landfill sites or as fertiliser derived from sewage sludge and applied to coppice being grown for biomass. The extraction of methane on a commercial basis is, however, feasible only from the largest landfill sites and will be unlikely to be an option on the more ordinary type of aggregate quarries. Methane can also be produced on a smaller scale from farm waste, such as animal slurry and litter, and from organic food waste from the processing and catering industries.

These matters are dealt with in more detail in chapters 4 and 5, but at this stage it becomes clear that there are real opportunities for changing the use of farmland and forestry, although some of these are being hampered by political and economic factors.

Harnessing the wind

The UK has an abundant supply of wind, especially in coastal and upland areas. This can be converted into electricity through turbines, although the supply so produced is intermittent and generally remote from the main points of demand.

To catch the maximum amount of wind, turbines are built in exposed positions and to a great height, so that they become highly visible in what are often sensitive areas of otherwise unspoiled landscapes, giving rise to planning problems. The procedures for establishing wind turbines on a particular site are explained in this chapter.

Wind is a natural feature of the countryside; and one that farmers may think of firstly as causing difficulties rather than creating opportunities. In the Fens, it may even lift a layer of topsoil, and seed, off the fields and blow it away through the neighbourhood. Elsewhere, the consequences tend to be less dramatic, but are still often destructive, with cereals being "lodged", lambs being weakened, barns being damaged and branches breaking off hedgerow trees and dropping onto public roads. How might this normally troublesome phenomenon be used to advantage, and especially financial advantage?

In the present context, the answer lies in being able to harness the wind into generating electricity and, technically, this could be either on a small individual scale or on a larger commercial basis. The latter offers the greater scope and is therefore considered first.

Commercial developments

The concept of converting wind into power has been widely promoted. Although controversial and not loved by everyone, the image of turbines delivering clean energy can be a compelling one, not just to environmentalists but also to landowners upon whose properties such turbines may need to be constructed.

As the country is committed to producing a growing proportion of its electricity from renewable sources and as most of this has so far been produced from land-based turbines, then it is fair to assume that there should be lucrative opportunities for landowners to diversify into what have already been given the rather rustic sounding name of wind farms. To be able to do so on a commercial basis will, however, depend upon a number of crucial practical issues.

The site needs essentially:

- To have a suitable flow of wind throughout the year.
- To have the topography to allow for a sufficient number of turbines.
- To be close enough to the local electricity grid.
- To be accessible for large construction equipment and materials.
- To be likely to gain planning permission.
- To be clear of local constraints, including housing, environmental designations and airfields.

A landowner will need to be aware of these factors before considering whether to embark upon a possible scheme and the main points are outlined later in this chapter. In most cases, however, they would be investigated in detail by the potential developers and it may be appropriate therefore first to mention the type of arrangements that are likely to be made by such developers when investigating a particular site.

Contracts

Potential sites may be identified either by the landowner or by a development company or by agents acting as intermediaries. In each case, once preliminary investigations have been made, the developer will seek to secure a series of agreements with the landowner. Embarking upon a potential wind farm project will involve both landowner and developer in a considerable commitment, even at the outset when carrying out feasibility studies and planning applications, let alone over the lifetime of the turbines once constructed.

The form of contract and option agreement under which this is done will therefore be of great importance. While most developers will have a preferred arrangement for this, any draft contract is likely to be adapted to suit the circumstances of each particular case and will probably need to involve professional advice. The first step may take the form of a letter of intent whereby the developer is given an exclusive right for a period of probably one year to carry out initial tests and enquiries, or "scoping". This will involve assessing wind speeds and examining the location and making preliminary planning searches, as mentioned below.

Assuming that these had produced sufficiently positive results, it would be followed by an option agreement for a specified period of generally between about three and five years. During this time, the developer would commit further resources to the project to secure funding, planning permission and grid connections. This agreement will then also allude to the final contract relating to the terms under which the turbines will be erected. Those terms will be incorporated within a lease for between 20 and 30 years and will need to cover a whole host of matters ranging from access and wayleave agreements to the form of payment and final decommissioning. That payment could be on the basis of an annual rental or a defined percentage of revenue, with or without a guaranteed minimum or participation in profits. The earlier short-term agreements are likely to be secured by a relatively modest cash payment to the landowner.

The purpose of these initial agreements is to provide a degree of commitment during the preliminary stages of the project to cover the considerable time and expense that is likely to be incurred. They will indicate the basic terms for the development itself, covering the main legal, financial and practical issues for the lifetime of the project, which are outlined below. The final contract will provide for a proper annual financial return to the landowner, while the first two stages tend to involve smaller sums.

Occasionally, the development will be open to community involvement, in the form of local people being able to invest in the project or to participate in a part grant-aided Energy Saving Scheme, which would both need to be covered by additional legal agreements.

Initiating a scheme

While it is possible for landowners to initiate wind farm projects for themselves, in most cases they will seek the involvement of specialist developers or consultants. The first approach to such organisations will need to be supported by some information about the basic site requirements listed at the beginning of this section. Of these, the question of wind speed may be the most difficult to envisage but will be one of the first that has to be answered. Surprisingly, this can be done relatively easily by reference to meteorological records that cover the whole country and can be accessed for any specified location. These will provide a preliminary indication to a developer as to whether a particular site is likely or not to have a sufficient wind flow to generate a commercial output of electricity.

The means by which this rather technical database can be used is explained in a later section, but also it may be helpful to consider the more fundamental aspects of wind as required for this basic assessment.

Wind scales

For a cluster of turbines to be commercially efficient they will need to receive a certain minimum flow of wind, in terms of both force and frequency. Even the most modern designs are limited as to the conditions under which they can operate effectively. Power can normally only be generated at wind speeds of no less than 10mph (about 4.5 meters per second) and no more than around 55mph (25mps), and tends to peak at around 33mph.

To visualise these parameters, it may be helpful to refer to the old Beaufort Scale. 10mph is 3 on the scale, with "leaves and small twigs in constant motion and the wind extending a small flag". 55mph is on the verge of gale force 10 and 33mph is at 7 or when "whole trees are in motion". These measurements are, by their description, taken at a few feet above the ground and allowance must be made for the fact that the turbine blades are positioned much higher than this, where there is likely to be a greater flow of wind than below. Nonetheless, on the days (and nights) when a cold mist lies motionless across the land or when, under a clear sky, the flag on the church tower hangs limp against its pole, the turbines will be inoperable. Even a "light breeze" as described by Sir Francis Beaufort will be insufficient. By the same count, on a day of scudding clouds and rattling windows, when wind seems to be in abundance, the turbines may have to be shut down. Some new designs of smaller turbines can now incorporate a mechanism whereby the blades are automatically adjusted when the wind rises allowing them to continue to operate at a lesser speed.

These restrictions are particularly crucial since, at present, there is no practical means whereby electricity can be stored, other than on a very limited scale in batteries. There is therefore no way of seeking to balance out the inactive days against those when the turbines are able to operate fully and this intermittency can cause operational and financial problems, as mentioned later. The proposed site must therefore be able to enjoy a suitable flow of wind throughout the year rather than benefit from occasional blasts of blustery weather, especially as one bears in mind that maximum output is reached at force 7 and that no advantage can be gained from times of stronger wind.

Intermittency

The wind power industry does take this intermittency into account and it is accepted that turbines can be expected to generate electricity at no more than about one-quarter of the capacity that would be produced were they able to operate constantly throughout the year. The commercial implications of this reduced output figure, or "load factor", is determined largely by the pricing mechanism for electricity in the UK and it contrasts with other countries, such as Germany and Denmark, where a more expensive tariff has enabled turbines to be built on sites with lighter wind patterns. There may have been a presumption, as mentioned in chapter 2, that in the UK the rising price of ROCs would in time enable a commercial return to be achieved in locations that might previously been considered to be inadequate. It has, however, now been suggested within the 2006 Government review *The Energy Challenge* that this form of financial aid should be reduced for the more commercially viable types of energy sources, such as onshore wind turbines, and focused instead on more innovative systems, such as offshore turbines and wave and tidal power.

The effectiveness of onshore turbines can be improved in some situations by increasing the height to which they are built. Average wind speed can be around 20% higher at 40 metres above ground level than at 10 metres, a phenomenon defined as wind shear. Commercial turbines that were previously placed on masts of about 30 metres are now being constructed to a height of over 100 metres, including correspondingly larger and more effective rotor blades. This may expose them to a better flow of wind, but it also raises two potential problems:

- The construction costs will increase and possibly negate the economic gain of the improved output.
- Being taller the turbines become more visible over a wider area and are more likely to encounter problems at the planning stage.

Wind speed

Having gained some indication of the level of wind at a particular location, it is then still necessary to observe it in terms of actual speed and the time for which it blows at various speeds during a period over which it can be fairly averaged such as a year. As mentioned earlier, it is possible to refer to meteorological records, and then make the premise that wind patterns experienced in a certain location in the past are likely to continue into the future. These records have even been transposed onto maps covering the whole of the country, so that it is possible to obtain an estimate of annual wind speed for any specific site.

This database, which is administered by the Department of Trade and Industry (DTI), has the acronym of NOABL and provides readings over areas comprising 1km^2 at a choice of three different heights above the ground. These readings show average speeds in metres per second, which may not be particularly meaningful in that format to a farmer or landowner but will give a useful indication to a consultant or other technician as to whether the site may be potentially viable. This basic information can be accessed by submitting an Ordnance Survey grid reference for a chosen location, probably most easily through one of the specialists in this field, as listed in appendix 1. There will also be local variations to these average readings that each cover a whole square kilometre, depending on topography and other physical features.

The next step, after obtaining a seemingly adequate result from the NOABL records, will be for a specialist to make a site inspection. This would be primarily to establish whether there are any potential anomalies within the immediate vicinity, such as hills and valleys, that could distort the readings for the wider area. This analysis will then usually need to be followed by a more specific assessment using an anemometer, as mentioned later. However, before taking this step one would need to have weighed up whether the proposed scheme was likely to encounter difficulties in meeting the other basic site criteria.

Accessibility

One of these criteria concerns the proximity of the site to a suitable power line. Electricity generated from the wind needs to be connected to the public network. Those sites that have a sufficient level of wind tend to be in remote and exposed areas that are some distance from any such power lines, which may furthermore be of only limited capacity. The construction of a link to the local grid by means of poles or pylons will again incur a cost that could be critical and raise potential planning problems. If the existing local lines need to be upgraded, further costs will arise, with the possibility of making the project unviable. The basic details of a local line can be checked early on with the electricity supplier for the area, by giving a map reference or quoting a number that can be found on most poles or pylons. A single phase supply (indicated by the presence of two rather than three lines) will invariably need to be reinforced before being connected to wind turbines. These issues are considered further in the section on grid connection on page 17.

These exposed sites tend also to be difficult to access, due to topography and to the distance from public highways, which must be substantial enough to carry construction traffic. Turbine sections are now being transported on trailers of around 30 metres long and over 3 metres in both width and height, which can be too big to manoeuvre through many country lanes. Given that the public roads are adequate, the cost of creating a suitable access across the landowner's property then needs to be budgeted for and may prove to be a limiting factor when appraising a relatively remote site.

Other natural features of the location will also have to be considered at this stage, such as the steepness of the site, that could be make construction work unduly difficult and costly, and whether it contains any springs that would be disturbed or contaminated by that work. The foundations for a single turbine of 100 metres high create a hole of around 400m^3, which is about the size of an olympic swimming pool. The potential for interference to springs and watercourses is therefore considerable, especially when constructing a cluster of such turbines. Furthermore, not only does the excavated material have to be transported from the site but the same volume of aggregate and concrete has to be tipped into it. The construction process will therefore need an adequate access road, of about 5 metres wide, not just to the edge of the site but also to the footings of each of the turbines, together with a permanent hard standing for cranes. All this, including one building to house the switchgear for the complete cluster, will use a substantial volume of building material, albeit not to the same depth as the foundation works.

Scale of development

The scale of the development should also be considered, in terms of how much land is likely to be required. To produce a viable return, a wind farm needs to be able to produce a commercial amount of electricity. Measured essentially in units of watts, this can then be defined in terms of a minimum number of turbines for a specific location and thereby the area of land required.

The amount of electricity likely to be produced by a particular type and size of turbine can be estimated according to the annual wind speeds across the site upon which it is being built. Recently constructed turbines have been rated at around 1 Megawatt (MW) each, whereas the newer and taller structures may be reckoned at up to three times that amount. To estimate the annual production, those figures have to be reduced by the load factor mentioned above. This was previously calculated on the basis of estimates made by the generators and reckoned to be at a figure of around 30%. However, it can now be more accurately assessed on the basis of the number of ROCs that have been issued in similar situations and tends to average about 25%. One turbine rated at, say, 1.5MW would be assessed as producing 25% of 1.5 or 0.375MW during the year which, when multiplied by the number of hours in a year (8,760) gives a figure of 3,285MWh. Such annual output is often recorded in kilowatthours (kWh), in this case 3,285,000kWh. The minimum output required from a development is determined according to the cost of construction, including the infrastructure mentioned above, and the anticipated return from the sale of electricity. The developer will then be able to assess the minimum number of turbines that will be needed for the project to be viable. This may range from about three to 20 or more, depending upon the situation, while in Scotland some schemes comprise between 30 and 40 turbines.

The scale of these larger developments has brought into question the way that they are still described as *wind farms* when they begin to look increasingly like industrial installations. They are now sometimes referred to as *wind parks* while a group of turbines may be more correctly described as a *cluster* rather than *farm*.

Construction costs may be measured in terms of the kilowatt capacity and will range around £700 per kW, which would equate to about £1m for each present day turbine. There will, however, be further costs incurred in providing ancillary works, such as building roads or transmission lines. The financial return depends on the price at which the electricity can be sold to the grid. This has been at around 4p per kWh, which would mean that in the preceding example an annual income of £131,400 might be achieved. The return to the landowner is generally linked to that income, either as a small percentage or in the form of an annual rental, which has tended latterly to be approximately £5,000 per turbine but can be significantly in excess of this according to circumstance.

Output

The output from turbines is measured in MW and generally referred to as the "installed capacity" of a cluster, meaning the amount that they could, in theory, produce if able to operate fully for 100% of the time. In practice, however, the actual annual production will average only around one-quarter of this.

It is also often quoted in terms of the number of households that could be supplied with electricity by them. This is a measure of the annual output of the turbines in MWhs when reckoned against the power requirements of an average household. This is rather misleading as it ignores the fact that wind-generated electricity is so intermittent and available for only about 25% of the time when consumers may need power. The claim, for example, that 10 turbines could supply 1000 homes conjures up an image of whole communities being unable to watch television or cook meals other than when the wind is blowing at the right speed!

In reality, the industry relies on a supply network that is still able to draw largely upon conventionally generated electricity while also utilising wind power as and when available. Perhaps it would be fairer to say that the 10 turbines could produce about a quarter of the needs of 4000 homes, rather than suggest that they could ever meet the full demands of any household, let alone those of an industrial or commercial enterprise.

Grid connection

The main purpose of commercial clusters of turbines is to sell, or "export", the electricity that they produce into the public system or grid. This raises a number of technical issues, particularly as regards fluctuations in supply. The output from even a single turbine of 6kW capacity or more has normally to be exported in total and cannot be drawn down at source and used on the property on which it has been generated. However, although landowners may not be able to plug into their own electricity, it may be accounted for as a rebate against the cost of using the public supply. This arrangement is likely to form part of the original contract with the developer and be separate from the rental equivalent being received on the actual annual output of the turbines. While switching electricity on and off seems such a simple matter when operated at home or even in a business environment, creating the flexibility to absorb wind-generated power can be complex and costly, especially in rural areas.

The supply of electricity needs to be constant, and any variations in output have to be anticipated and regulated. This is particularly so in rural areas where most clusters are located and where supplies tend to be transmitted in lower voltages and spread out over large distances to relatively isolated consumers. Even during a period of settled weather, there will be constant variations in the strength and turbulence of the wind, which then cause fluctuating output from the turbines. These become extreme when the wind rises to a level at which the turbines shut themselves down or, in some designs, when they then restart as it falls again to a manageable speed. The method, and cost, of absorbing this into the public supply depends upon such circumstances as location, the capacity of the local network, and the size, number and design of the turbines. The solution may involve upgrading the local supply lines or connecting to a more distant grid with a larger capacity. Each will involve significant costs that need to be clarified in advance with the local supplier. In the case of installing new transmission lines, the cost may have to include that of agreeing wayleaves over the intervening land. They will also need planning permission, which can be as difficult to achieve as that for the wind farm, especially where they pass through otherwise unspoiled countryside.

Wind farms are generally connected to a local grid and are not linked directly to the national network. However, National Grid, the company responsible for this network, has demanded that developers provide bank guarantees towards the anticipated cost of expansion. This is to provide National Grid with some protection against the possibility that it may have had to invest in improvements to absorb the output from proposed wind farms that then fail to materialise. While this would not involve individual landowners directly, it is likely to put an additional financial burden on developers and to affect their ability to promote new schemes.

Site area

Each turbine has to have a certain amount of space around it in order to operate effectively and without being impeded by turbulence, or "shadowing", from others. This may depend on the particular terrain and situation but, as a general rule, they need to be at least 300 metres apart. On an even and open site a cluster of, say, 10 turbines may occupy an area of about 100ha (or 250 acres). The structures themselves and the access tracks take up only about 1% of this so that most of the land remains available for farming. Where the site includes uneven ground, such as ravines or gullies, then the spacing may well need to be wider to allow for constructional limitations, as well as the possibility of uneven wind patterns. During the construction process, which may take several months, a much larger part of the site than just 1% will be needed as a working area and farming may have to be abandoned there during that time. The ensuing loss of income and the temporary damage to the land surrounding the actual structures may be compensated under a supplementary agreement with the contractors.

The locational issues mentioned so far are essentially of a physical nature, concerning practical features, such as wind speed or access and their economic implications. There is one further matter that can have a major impact on the feasibility of a proposed development but which is more difficult to define and predict, namely planning.

Planning – initial considerations

No new construction or change of use can take place within the country without the consent of the local planning authority. The procedures for this are dealt with in a later section, but there are some general issues to be considered at the outset and before getting to the stage of making any formal application. Such applications are all considered on their individual merits, although there are established policy guidelines in all planning matters that need to be adhered to and that create the parameters within which the local authority and ultimately the national Government have to operate. The main such criteria for proposed developments of wind turbines are:

- Conservation.
- Proximity of roads and houses.
- Disturbance to air traffic.
- Construction of new transmission lines.

Exceptions may be made to the criteria that have been laid down for these particular issues, especially when a case is well argued or there is a strong desire among the authorities to allow the proposed scheme to proceed. It would be as well to make a general assessment of their potential significance at an early stage and prior to embarking on detailed submissions.

Conservation

Places within Britain that are likely to have a sufficient wind flow for the commercial production of electricity are inevitably in exposed and highly visible locations. These are often also in beautiful and highly cherished landscapes. Any proposal to erect a cluster of turbines along the skylines of such areas is bound to excite controversy and opposition. Applying for planning permission under such circumstances can then be an uncertain and costly venture and is proving to be a significant constraint on the further development of onshore wind power. Changes in national legislation are being implemented to make it easier for local planning authorities to grant permission for new wind farms, but such projects can still meet with considerable difficulties. Many of the sites that would be deemed to have an appropriate wind pattern are in areas with special designations, such as being of Outstanding Natural Beauty or National Parks. Those areas that are so defined have been granted special statutory protection against most forms of development, the implications of which are considered in a later section.

Such formal designations are not only for the protection of present day landscapes and environment, but also with the past in the form of ancient monuments. Hill tops and ridgeways, which can have the right physical characteristics for wind power, have been utilised over several millennia for fortified settlements and thoroughfares and are now protected as such for historical and archaeological reasons. Designations of all kinds are recorded on maps by the Local Planning Authorities and it can be easily checked at an early stage as to whether a proposed site lies within one of these areas and whether, therefore, it might fail to be granted permission.

Other sites may not be so formally defined but will still be seen by the public as being unsuited for this kind of construction and therefore attract strong opposition. Some areas of uplands or coastline that are not at present designated and which would appear to have a potential wind supply may well be earmarked for designation in the future. There is a growing commitment among governments within the EU to increase the areas in which protection is to be given to environmental features such as habitats and wildlife. It is possible then that planning permission for a proposed wind farm development on a wild, open and undesignated site would have to be withheld due to the fact that the location was scheduled for inclusion in a conservation area in the future. Where these possible future sites may be could be difficult to establish, but it is easy to investigate whether the proposed development would lie within existing designations and so be likely to encounter serious constraints if a planning application were to be made.

On the other hand, as these designated areas have been protected from development, land that lies immediately outside their boundaries becomes a more attractive proposition as it may still benefit from similar topography and so also wind patterns but escape the tight strictures of the planning regulations. These issues are considered in more detail in the later section on planning permission.

Another element of conservation that is frequently cited when new wind turbines are being proposed is that of bird migration. There has been growing evidence and concern in the UK and elsewhere about birds being killed by flying into turbine blades. This is particularly so for clusters built in coastal areas, which are chosen for their favourable wind conditions but which are also often on routes used by migratory birds. It can also arise on open moorland that provides habitat to certain endangered species. In such cases, planning permission may only be granted if the turbines are sited in a format that avoids the most frequented flight lines or on a condition that they are switched off during the main migratory season. That can have commercial consequences, either through the loss of income from being unable to operate the turbines for part of the year or by having to modify the layout and thereby possibly lose some of the best wind flow.

Although most countrymen will know and appreciate the wildlife on their land, they may not be aware of the significance or actual location of the migratory routes. It can however be readily ascertained by reference to possibly the local authority and certainly the Royal Society for the Protection of Birds (RSPB). Where the proposed development is along the coast or in an area known for its rare moorland birds, it would be worth making enquiries prior to the full design and planning stage. Doing so may, however, cause another difficulty in that it alerts the public to the fact that a wind farm is being considered. The implications of this are considered in the section on wind monitoring.

Roads and houses

There have been concerns about the effect that turbines may have when built near to houses, both as to the welfare of the residents and the value of the properties. In response to this, an industry guideline suggests that turbines should not be sited within about 500 metres of a house, where that household is not financially involved in the development. If it can be seen that there is housing close to a proposed site, then it will probably be necessary to redesign the layout of the cluster before it is submitted to the planning authority.

It is also recommended within the government planning guidelines (PPS22) that the minimum distance that turbines may be to a public road should be equivalent to not less than the height of the turbine, now generally at least 100 metres. This is presumably to reduce the visual distraction and possible downdrafts that they might cause if built too close to a road.

These locational constraints can best be assessed by drawing on a map a circle of about 500 metres around each house and a parallel line 100 metres or more in from each side of a road. It will then need to be established with the developers whether a viable cluster of turbines could still be built in the areas remaining outside these lines.

Air traffic

Another feature that is worth checking at an early stage is whether development on the proposed site is likely to interfere with radar and telecommunications installations. Due to the height and movement of the rotor blades, wind turbines can affect radar and other signals and the Ministry of Defence (MoD) and the Civil Aviation Authority both have stringent requirements that can prevent them being constructed within a certain radius. This can affect a surprisingly wide range of places, since the MoD requires that a distance of not less than 45 miles be kept clear around its installations, which is far greater than in most other countries, such as in Germany where the equivalent figure is only 5km, or about 3 miles. Indeed, on this basis, the MoD has raised valid objections to about half the total applications made for wind farms. The likelihood of whether this issue might arise for any specific location can be checked on the internet or through consultants who specialise in this field.

There has been a review of how wide the radius need be within which turbines are deemed to interfere with radar. If this could be reduced, new sites that were previously excluded might become viable. There are apparently two possible solutions that would be acceptable to the MoD: the development of new software and the the installation of an electronic tilting system on each turbine. Both require considerable time and money and seem unlikely to be pursued in the near future. Indeed, the problems of radar interference and increasingly also of low-flying aircraft may become greater with the increased heights of turbines.

Residential areas

Although most sites suitable for wind turbines are in remote and exposed locations, it is also always possible that there may be some housing within the vicinity. This could become increasingly the case as developers need to consider a wider range of potential sites when others have already been taken or irrevocably rejected.

The height and visibility of the masts can mean that their potential impact upon residential interests could cover a relatively wide field. Concerns from local people can go beyond the issue of visual intrusion upon their neighbourhood and now encompass questions of health and disturbance. This relates mostly to the noise that is emitted by the turbines the flickering effect that can be experienced when in constant view of these structures. Other potential problems may arise in disrupted reception of television and mobile telephone signals. While such matters are not normally as critical to the proposed development of a cluster of turbines as others already mentioned, such as wind speed and designations, it could become significant at the planning stage and will need to be identified at the outset.

Wind monitoring

When it has been established that the proposed site is outside any area of designation or radar transmission and is sufficiently accessible both to the electricity grid as well as to public roads, while unlikely to dominate residential property, the evaluation will revisit the question of wind speed. It is unlikely that the NOABL records will provide sufficiently specific evidence on which to launch a long-term investment in a wind farm and the readings will need to be checked on site. This can be done by erecting on the site itself an anemometer, which comprises a small propeller with a vane that is connected to measuring equipment. Although that would seem to be a relatively easy step to take, just putting up a light piece of equipment for a temporary period, it does tend to mark the opening of a pandora's box of potential problems. Initially, these will be the matter of cost and then of publicity.

The installation of an anemometer can cost around £10,000 as well as further commitments that are referred to below. It is probable then that the developer will, at this stage, need to draw up some formal agreement with the landowner before embarking upon these situations. Meanwhile, there is a more practical issue to be resolved.

An anemometer needs to be as high as possible, so as to be able to monitor wind speeds at around the height at which the turbines will be built or to provide accurate assessments of wind shear. It also has to be in place for a long enough time in which to gain a reliable measure of the wind patterns in that area, and not be distorted by outbreaks of untypical weather, such as recurring storms or unusual periods of calm. This is likely to be between 6 and 12 months. In some situations, it may even be necessary to have more than one anemometer in order to cover a number of positions within one general site.

Under current legislation, any structure of more than 12 metres high that is to be erected and then left on site for more than 28 days will require planning permission. This has a number of implications and it may be feasible for a developer to postpone those particular problems by working at first with readings taken nearer to the ground and over a shorter time span.

Applying for permission costs time and often also money but, in this context, there is a further, more crucial, consequence in that the application will bring the proposal into the public domain. Planning applications are recorded on the internet and local newspapers as a means of keeping the neighbourhood informed of what is being proposed in the vicinity. The landowner is also required to display an official notice on the property giving a brief description of the scheme. The erection of an anemometer will be recognised as a prelude to the development of a wind farm and the announcement of the application will alert local and other interests. In the majority of cases, this leads to a campaign of opposition, which can arise in many types of planning applications but, in this instance, it does so at an unreasonably early stage.

To be effective, opposition to a project such as a wind farm needs to be based upon technical planning issues rather than emotive judgments, but the latter do inevitably arise and can be levelled at the applicant or landowner. Therefore, in taking what appears to be just a preliminary step of applying for planning permission for a temporary anemometer, landowners may well be putting themselves into the firing line, which they will need to endure for the whole term of the planning process, which can last several years.

The local community and environment

The procedure for gaining planning permission for a wind farm is likely to require a significant commitment in both time and money and will tend to be organised by the developer who will, in turn, call in other professional experts, such as environmental and scientific consultants and lawyers. However, there are some important issues that can be addressed by the landowner at the outset.

The first may be to take the initiative to introduce the matter to the local community and to explain what is likely to be involved, both in terms of the physical development being proposed and also the benefits expected to accrue from it. These benefits will probably be couched mainly in terms of protecting the global environment by contributing to the overall reduction of carbon emissions. There may also be opportunities for local people to invest in the scheme and thereby gain some financial return from it. It is difficult for the community to gain a practical return in the form of being supplied by their own "green" electricity, which might have a certain appeal. This is due to the complications that arise in connecting a fluctuating source of generated electricity to the local grid. Since wind power can only be intermittent, according to weather conditions, the grid has to be able to supply conventionally produced electricity to the users around the wind farm at all times, regardless of whether the turbines are engaged. It would be too complex and costly to incorporate a system that switched in and out of the grid according to the performance of the turbines.

On the other hand, developers are increasingly able to offer an annual payment towards community projects. This might appear to be little more than a blatant bribe or sweetener to overcome objections, but it does reflect a widely used principle in Town and Country Planning where developers can be formally required to fund some public work, like the construction of a village by-pass, as a condition of being granted permission for a major project, such as a new housing estate. Indeed, it has been proposed in Scotland that a statutory minimum level of community payment should be imposed upon wind farm developers. Latterly, an annual contribution of around £1000 per MW might be offered as part of the development package. Depending on circumstances and location, this may be part-funded by the EU Energy Saving Programme.

Local interests

Such matters might best be raised in discussions with the Parish Council and in open meetings with the neighbourhood residents and the local press. In many cases, these local interests do gel into wider groups of well-organised campaigners, who may be more difficult to address informally but who will have sincere concerns that will need to be dealt with. These may include specific pressure groups committed to the protection of particular countryside areas or features. If the landowner is to maintain a dialogue with such organisations, rather than a confrontation, it will be important to recognise the value of compromise. A preservation society may be receptive to discussions about adapting specific sites for the turbines, so as to cause a lesser environmental impact, and should certainly be ready to learn about the commercial and other restraints that would make such proposals difficult or unworkable. Others may have a more particular agenda, such as the RSPB seeking to protect migrating birds from flying into the turbine blades. Discussions over possibly re-siting some of the turbines or restricting their use at vulnerable times could avoid confrontation at a later stage. These various organisations are entitled to objections or suggestions to the relevant planning body and are generally well equipped to do so. Their cooperation, or at least understanding, can be important once the whole matter goes to the crucial stage of a planning application.

The policies of the local authorities on renewable energy will also be highly important and will tend to have been checked by the developer at an early stage. These include not only the Parish, District and County Councils, but also the Regional Development Agencies and the devolved administrations of Scotland, Wales and Northern Ireland, all of whom will have taken an official view on the principles of renewable energy.

Larger schemes, where output is expected to exceed 50MW, are referred to the DTI in London, or to the Scottish Executive for those located in Scotland.

Planning permission

The issue of planning permission has already been flagged as a potential problem involving time, money and publicity and it is matter that will normally be dealt with by the developers as they have the appropriate experience and resources for it. That there will be difficulties is certainly likely to be the case in any major project, but planning does not necessarily need to be an adversarial process, especially at the outset. There is, for example, every opportunity to take informal soundings about a proposed development before making any formal application.

The front line in this process is the Planning Office at the local district or borough council, whose staff will be able to report on existing policy and recent practice and also advise on what features may need to be incorporated in order to increase the chances of success. In some metropolitan areas, the relevant body will be the unitary authority. In most conventional cases of building development, reference would first of all be made to the Local and Structure Plans. These have been drawn up every few years following a consultation process, initiated by the County Council in the case of Structure Plans, who also have responsibility for strategic matters that could include renewable energy. This system of locally agreed Plans is being reviewed and due to be replaced by Spatial Strategies together with Local Development Frameworks.

The County Councils are directly engaged in local planning in Wales and within National Parks, while in Northern Ireland this is handled by regional offices of the Assembly. The prevailing Plans set out the overall agreed planning policy for the area, notably showing zones that have been earmarked for various uses, including development. These existing Plans will not have defined any sites specifically for the development of wind farms and the sort of location that would be physically suited for such a purpose will be defined on the Plan as being for existing, agricultural use. It would also show those areas that have been designated as being of some particular conservation or landscape value.

Latterly, especially in the devolved regions of the UK, certain areas are being defined as being Strategic Search Areas (SSAs). These are in the form of maps showing locations that are not protected through designation but which may have suitable wind conditions for turbines. This does not imply that all other factors may be found there as well, such as viable access for roads and grid connections, but it does suggest that the process for gaining planning permission may be easier than elsewhere. An example of this policy is found within Wales where there are a series of areas referred to as TAN 8 sites. These derive from Technical Advice Note 8, which was produced by the Welsh Assembly and which has had the effect of attracting special interest from developers to locations defined by it.

Planning – environmental issues

It will be probably at this stage that it can be established as to whether or not an Environmental Statement will be needed in support of the planning application. Most planning authorities will require that an Environmental Impact Assessment be prepared for any scheme over 50MW. This is a matter that would normally be dealt with by the developers, having to engage consultants to carry out the work when appropriate, but it does have implications for the timing and cost of the process, and does put an additional onus upon the developers. Schemes of over 50MW are furthermore required to be put to a public enquiry, under Section 36 of the Electricity Act 1989.

An Environmental Statement will also be an essential part of any proposal to erect wind turbines in a protected designated area, or on an adjacent site that is likely to have significant environmental effects on such an area. These areas comprise:

- National Parks.
- Areas of Outstanding Natural Beauty.
- Sites of Special Scientific Interest.
- Environmentally Sensitive Areas.
- The Heritage Coast.

In Scotland, the same principle applies, but some of the sites are defined differently, as National Scenic Areas and National Heritage Areas. While in the past it has proved to be impossible to secure permission to develop wind turbines within these areas, there is now a means of at least making one's case for it, initially by submitting an Environmental Statement. Some authorities are prepared to consider applications where it can be argued that the environmental benefit of the proposal outweighs the harm. A few have even commissioned their own Landscape Character Assessments in order to establish whether there might be sites within designated areas where the likely environmental impact may be of lesser consequence.

The balance between benefit and harm is difficult to find, as there are such divergent concepts involved; the global benefit of reducing carbon emissions cannot be measured specifically against the local harm being done to a beautiful landscape. Not only are the basic premises widely divergent, but also individual opinions vary too; as to whether global warming is scientifically proven, whether any carbon savings are achieved through the use of wind power, and whether turbines actually damage or even enhance the countryside around them.

These issues are often emotive and involve difficult areas of politics and commerce as well as sincerely held views, but they have to be seen in the context of normal planning practice. The system is generally very protective of rural landscapes, especially those with special designations. Detailed controls are imposed on any new development, even down to the colour of roof cladding that may be permitted on a new lean-to against a farm barn, and yet, in the same context, developers are expecting to be allowed to erect a cluster of industrial structures rising 100 metres high.

An example of how divergent planning rulings can be is shown by a case at Lambrigg in Cumbria where, in July 2002, a farmer was required to demolish an extension that had been built onto the rear of the farmhouse without planning permission. The irony of the situation was that the property and the surrounding landscape had been dwarfed by five wind turbines, which made a greater impact on the locality than the domestic extension that was hidden from public view.

Planning procedures

Applications for routine or smaller-scale developments are generally assessed by the Planning Office acting on its own authority, but those proposals that are of a larger or more contentious nature are likely to be put to a Planning Committee for a decision. That committee is composed of elected councillors, each of whom may act upon their own interpretation of the case, particularly as to how it conforms to the Local and Structure Plans. There is opportunity to lobby the councillors, and an applicant, such as a landowner, may be better placed to do so initially than the developer, having local connections and know-how.

There is opportunity to discuss a proposed scheme with the Planning Office prior to submitting a formal application. This may enable one to modify certain features of the scheme so as to make it more acceptable to the Authority. It is not uncommon also for permissions to be granted to developers subject to their agreeing to provide or contribute to specified facilities within the local community. These are referred to as Section 106 agreements and will involve the developer in additional costs and often also further delays.

If the application is refused, whether by the Officer or a majority of the Committee, an applicant may resubmit it in a suitably amended form but alternatively has the right to appeal to the relevant government agency. The appropriate Secretary of State will be expected to submit the matter to appeal or inquiry. This will lead to a formal hearing, at which the case made by both sides is reviewed. The format of such hearings varies according to the different devolved regions of the country, but the decision arrived at by the chairman is binding upon all parties and can only be disputed on a point of law. Appeals may already have involved legal advisers, as well as specialist witnesses, such as environmentalists and chartered surveyors.

The national Government does not otherwise have any direct control over the local planning authorities and its views and policies have, until recently, been presented to local councils in the form of Planning Guidance Notes, which, as the name implies, are essentially only directives as to what the Government would wish to see implemented. These have now been changed to Planning Policy Statements (PPS) and are intended to bring a closer degree of direction over local authorities in such matters. Energy issues in England are dealt with particularly in PPS22. In Wales, this is currently covered by Technical Advice Note 8 (TAN8) and in Scotland this has been under National Planning Policy Guidance 6 (NPPG6), which is being reviewed within Scottish Planning Policy. (SPP6).

In England, council policies may also be influenced by the stance taken by the Regional Development

Agencies, especially in such matters as renewable energy for which there may be local targets. In the other regions of the UK, much of the planning policy is determined by the devolved bodies, namely the Scottish Executive, the Welsh Assembly and the Northern Ireland Executive. Where these have particular leanings towards wind power or have adopted specific targets for renewable energy, local planning decisions may be overruled by the regional body, even after the Appeal process.

According to intentions stated in the 2006 Energy Review, the system of planning inquiries is due to be reviewed, with the purpose of reducing the delays that have been incurred in hearing planning submissions for renewable energy projects. The Review states that it has been taking an average of 21 months for consent to be granted for the development of wind farms. (It can then take on average another 20 months between gaining consent and being able to commence operating, due to delays in connecting to the grid.)

While the planning applications themselves will normally be made by the developers, the landowner is bound to become involved in such matters as establishing links with local councillors and pressure groups and perhaps also in presenting the concept to the local press and community.

Property values and taxation

The Royal Institution of Chartered Surveyors conducted a survey in 2004 to consider the effect of wind turbines on property values. This was inevitably constrained by the limited number of developments upon which to base the assessment and its conclusions are therefore rather tentative. The majority view was that the construction of a cluster of turbines would be likely to impact upon the values of residential premises within the locality but that there was no discernible effect on the value of the farmland itself. If there were in due course to be cases where neighbouring house owners could prove that their properties would be devalued as a result of a proposed development, it could raise questions of compensation. For the time being too, it seems that there is little evidence to suggest that there would be any significant enhancement in the value of the land upon which the wind farm has been erected, despite the fact that the owner will benefit from a new income stream once it has become operational.

The development of a cluster of wind turbines could also affect the landowner's tax position, in that the area of land underneath the turbines may no longer be accepted as being agricultural, so its value would then no longer be eligible for agricultural property relief against inheritance tax.

Business rates are now applied to newly constructed wind turbines and can involve a significant annual cost. They are calculated according to an assumed capital figure per MW of "installed capacity", although there have been proposals also to assess them against the values of the turbines themselves. The actual amounts will need to be checked for a specific local authority at the time, but the rate levied has tended to be a little over £2000 per MW in England and Scotland and around £800 in Wales. These rates would not be chargeable if all the electricity were consumed on the property and none were sold into the grid.

Single turbines

The emphasis of wind energy development within the UK has been largely on the construction of clusters of several turbines, rather than single structures. This is due to the fact that the costs incurred in initiating these projects, such as in gaining planning permission and negotiating connections to the grid, can be almost as great for a small scheme as for a larger commercial one, even though there is a lesser financial return for the former. Smaller-scale generation is now receiving some Government support, in the form of grants under the Low Carbon Buildings Programme (LCBP) and is confirmed as a matter of policy in the 2006 Energy Review in urging the implementation of a Microgeneration Strategy. Individual grants are capped at £5000 and paid only on the first 5kWe generated. The LCBP replaces the Clear Skies Programme and is part of the Communities Renewables Initiative (or Scottish Communities and Householder Renewables Initiative).

There are occasions when a single turbine may be constructed alongside an industrial or office building, but there may be reasons for this that may not apply so readily to a farm or estate. The consumption of electricity in a commercial enterprise will be greater than that on a farm or estate, which may make the exercise more viable. There might also be a further justification in that it may be corporate policy to actively embrace environmental measures or that it may have been a condition of gaining planning permission for the building itself. A commercial consumer of electricity, such as an industrial complex, may have another incentive to erect a wind turbine in that it should then be able to gain exemptions against the CCL, which is being charged on business users of "mains" electricity. In the rural context, single turbines can be considered in certain situations, either as a stand-alone system or when connected to the grid.

The cost of installing a single turbine varies considerably according to circumstance, particularly as to location and grid connections. One that would be capable of servicing a large farm, requiring an output of around 6kW, may involve a gross sum of around £20,000 which could be offset by an LCBP grant as mentioned above. Figures from the DTI suggest that such a unit would produce annually about 10,000kWh, with a saving of around £700 at prevailing prices, requiring a payback period of 29 years. Microgeneration of electricity may therefore be more of environmental than commercial interest until such time as the technology improves and energy prices rise.

Stand-alone systems

The limitations of generating electricity for one's own consumption alone is that it depends upon being able to store power for such times as the turbine is not working when the wind speeds are either inadequate or excessive. Storage still has to be by means of batteries, which have limited capacity and become unduly costly if that capacity were to be raised. The concept of using an exposed and possibly remote location to generate enough electricity so as to be independent of the public supply is likely to be feasible at only about 50W and would therefore be limited to the needs of a single off-lying building.

Grid-connected systems

A single turbine on a farm can be connected to the grid depending upon its output and upon the loads already being carried by that grid. This will tend to be appropriate only to the larger turbine, but even then there could still be a disproportionate connection cost compared with a cluster of several such turbines. The case for constructing a single turbine may depend on environmental considerations rather than financial factors, which, in turn, are likely to hinge upon the future market price of electricity.

It may be more effective to erect a single turbine on a farm where it can be freestanding than on the roof of an urban house. But there are limitations as to the distance over which this relatively small output can be cabled. For technical reasons, most micro-systems feed the generated output into the local mains and the property continues to use the normal public supply, albeit on a tariff that recognises the level of electricity delivered from the site and the value of ROCs associated with it. This means that if the grid fails, through storm damage or another cause, the property will be without electricity as the turbine will automatically shut down until normal power is restored.

Action points for farmers and landowners

- Check whether the site lies within special planning designations. If it does, meet with the local planning officer to discuss the general situation, although gaining permission will be difficult or impossible.
- Ascertain average wind speeds via internet enquiry and quoting the grid reference to consultants or developers.
- Check the location and status of the local electricity grid via an inspection and reference to a supplier.
- Consider matters of tenure, such as the position of farm tenants and arranging wayleaves over neighbouring land.
- Consider the question of access for construction and maintenance and its effect on neighbouring properties.
- Instigate initial discussions with consultants or developers as to the feasibility and longer-term financial and other commitments.
- Weigh up the potential reaction of the local community and other interested parties, and the likely impact on the landscape, farming business and sporting opportunities during construction and over the longer term.
- Assess the implications of business rates and of proposed changes to the Renewables Obligation on the viability of future projects.

Biomass, in the context of renewable energy, refers to crops and residues that are used as fuel to produce heat and power. It can also involve the processing of organic waste into methane gas.

This chapter explores the materials that can be manufactured into fuel, which comprise primarily cereal straw and coppice and other timber by-products, as well as methane production, which is based mainly on manures from dairy, pig and poultry farms.

Biomass

The concept of generating energy from ingredients that are already being produced in the countryside is a compelling one, and it seems indeed to embody the whole principle of renewables in that it uses crops that can be regrown once harvested and waste materials that are also part of an ongoing cycle. Crops that are specifically cultivated for this purpose are sometimes described as "dedicated" biomass, while waste materials and by-products, which include cereal straw and timber residues, are defined as "dependent". Coppice forms the main type of biomass crop in Britain, either as woody coppice or as miscanthus.

Woody coppice was originally grown under woodland trees, particularly oak, and comprised species such as hazel and chestnut. These plants, having been rooted within the wood were cut regularly at a semi-mature stage to produce small-sized timber that could be used for the manufacture of fences or joinery or burnt as fuel, either as wood or after having been converted into charcoal. The plants have an ability to regrow from the original root stock, or "stool", and can then harvested again over a cycle of a few years. Evidence of such coppice can still be seen in broadleaf woods across England, but it is now only rarely being managed as such, due to limited markets and lack of viability.

It can also be grown on open farmland as "Short Rotation Coppice" (SRC) using fast-growing tree species, notably willow, or occasionally poplar. In this case, the coppice is planted as small sticks, which then form roots and grow within three years to produce saplings that can be harvested and ground into woodchips. The cut stems that remain in the ground then put on new growth and repeat the cycle.

Miscanthus is another form of coppice that can be grown on farmland in this country and then harvested even on an annual basis. Although not a woody plant but more like a tropical grass, it produces a material than can be processed into the equivalent of a wood fuel as well as other products, such as equine bedding.

Where these forms of coppiced biomass are used to supply power stations, it is possible to generate electricity, and heat, by a means that does not draw further upon the finite resources of fossil fuels and which is in itself totally renewable. In that respect, biomass is able to fulfil one important part of prevailing environmental policy, namely reducing the use of fossil fuels. However, unlike wind power, which, once established, is virtually free of any emissions, biomass conversion involves the process of burning and thereby creates smoke containing carbon dioxide. It is as well therefore to consider whether such processes can make any positive contribution to the Climate Change Programme and the targets for reducing carbon emissions.

Climate change

Plants, whether in the form of trees or cereals, absorb carbon from the atmosphere while growing and retain it until such time as they are consumed or decompose. When timber from trees is burnt, the retained carbon is released as carbon dioxide, but this can be balanced by the amount of carbon being constantly absorbed by other trees that are being grown to replace those that were destroyed.

In an environment of sustainable forestry, such as that practised in the UK, this balance is secured by the obligation on foresters to replant any woodland areas where trees have been felled. Similarly in the case of coppice, crops that are harvested automatically produce new growth in place of that which was removed. The equation is not quite equal, in that other forms of energy are being expended in the harvesting and processing of these fuels, mostly in the form of oil with its resultant emissions.

Emissions

Wood is generally considered to be "carbon neutral", implying that on an acre-to-acre basis timber or coppice production basically absorbs as much carbon dioxide as is released when it is processed as fuel. If timber were left to rot and decompose, as does occur in some nature reserves and indeed now also within commercial silvicultural systems, the carbon that had been fixed within those trees will be released into the atmosphere. This is mostly in the form of methane, which happens to be a more damaging form of "greenhouse" gas than carbon dioxide.

Where biomass of whatever kind is burnt for power and heat, the steam and smoke that arises is subjected to stringent treatment in the exit flues. For example, the emissions from straw being processed in a power plant are far more controlled and benign than the great clouds that used to blanket the countryside when it was simply set alight in the open fields!

Figures for the efficiency of biomass in terms of carbon emissions are still a little imprecise, but estimates suggest that the use of such fuels can result in a 90% reduction against the same heat or power being produced by more conventional fossil fuels. The balance between carbon emitted by wood when being burnt as fuel and that which is stored or sequestered during the growth of the timber or coppice is difficult to quantify and the use of wood fuels is not as yet accepted as being an eligible process within the Emissions Trading Scheme (ETS).

Nonetheless, biomass is able to make a positive contribution towards to the agreed targets for a reduction in carbon emissions. It also has two further advantages within the present rural climate:

- Introducing new uses for farmland and woodland at a time when both agriculture and forestry are under financial pressure and in decline.
- Creating potential conservation habitats.

Moreover, in contrast to wind power, biomass can generate electricity (and heat) whenever demanded, thereby avoiding the need for a constant back-up facility. Yet, despite these apparent attractions, only a small amount of heat and power is currently being generated by this means, which seems due largely to the cost of developing the necessary generating plants. Those costs (estimated at around £30m for an output of 10MWe) are modest when compared with the development of, say, a nuclear plant but they have none the less been difficult to fund at a time when consumer prices for electricity were relatively low and without more specific government support.

Location

Another issue with biomass is that the material used for firing the power plants is relatively bulky and thus costly to transport over large distances. Therefore, the industry depends upon being able to establish local plants within areas where coppice and other products can be grown. This issue of proximity is considered to be sufficiently critical that grant aid, under the Energy Crops Scheme (ECS), (or the Farm Enterprise Grant Scheme in Wales) for the growing of SRC or miscanthus as an energy crop is available only if the applicant has a guaranteed end-use for the product. This may be in a power station or to a community energy scheme or even within a heating plant for individual consumption.

The guidelines given by the Department for Food and Rural Affairs (Defra) suggest that the crops should be grown within 10 miles of smaller installations and 25 miles of larger units. In practice, however, there may be situations where newly established power stations are able to pay commercial rates for material brought in over greater distances than these, particularly while more local sources are still being established. It should also be mentioned that the conditions of grant do allow for growers to work through approved supply groups or "middle men" as an alternative to being contracted to one specific installation and this can give some flexibility to the question of location. There are developments too in the process of storing and compacting coppiced material, which can lower the volume of the delivered crop relative to its weight. Wood fuel is now also being converted into pellets and the processing plants for this may provide new locational outlets for biomass producers.

There are some climatic limitations as to the locations in which coppice may be grown successfully. Originally, miscanthus was considered to be best suited to the relatively mild area of south west England, although it is now also being grown as far north as Yorkshire. Woody coppice, notably willow, prefers wet conditions, which might suggest that it is more likely to be seen in the western half of the country. It is also grown in the drier arable areas of the Midlands and eastern counties, especially where it can be established on wetter ground beside watercourses.

While the development of biomass as a fuel may be hampered by the fact that current government policies favour electricity rather than other forms of power such as heating, it should be recognised that there may be fewer ongoing costs in transmitting electricity to where it is needed than in transporting woodfuel to existing power stations.

Government policy and grant aid

The use of biomass as a source of energy is still very limited, suggesting that it is not yet viable in commercial terms. This may be due to the time and costs involved in research and development and the fact that the process is less well suited to generating electricity than to producing heat. Unlike electricity, which can be fed into an existing network, heat needs to be channelled through newly built systems, which require further investment. Without this, one would be left with just an electricity supply that is relatively inefficient and therefore uncompetitive. One solution, as mentioned in chapter 2, is to supply biomass into specially built CHP units that use the heat initially to drive turbines to generate electricity and then channel it into community heating systems. If biomass is to be successfully utilised, there needs to be a financial position from which the necessary infrastructures can be built and the feedstocks also then cultivated. This inevitably relies on state funding and therefore on government policy.

In the rural sector, biomass has been supported primarily through the ECS, which provided capital grants for the establishment of SRC and miscanthus and for setting up producer groups. At £1000 per hectare for coppice and £920 for miscanthus, these were assessed according to an estimate of standard costs of establishment, with 50% being available for woody coppice and 40% for miscanthus. A higher rate (£1600 per hectare) was paid for coppice being grown on any land that was formerly given over to livestock. This enhanced payment was intended to reflect the loss of income from the former livestock enterprise. The lesser rate for miscanthus was due to a limit imposed under the Rural Development Regulation, rather than implying that there are lesser costs or benefits arising from this particular form of coppice.

Grant aid for producer groups was on a sliding scale, starting with the possibility of 100% funding of specified costs in the first year. The ECS excluded support for producers of fuel from miscanthus, which came instead under the Bio-energy Infrastructure Scheme (BIS). This also included mention of other similar grasses, such as switch and reed canary, and was part funded by the Lottery. The ECS was part of the England Rural Development Programme (ERDP) that closed in 2006 and is due to be replaced by a new programme for the period 2007-13. At the time of writing, this was still awaiting confirmation from the European Commission and it remains to be seen whether the aid that had been available under the earlier ERDP will be offered again under the new programme. A similar situation applies to the equivalent schemes as administered in Scotland and Wales. It is possible that any new scheme for energy crop establishment could be at lesser rates of grant than those that were available previously. Farmland, including setaside, that is used for growing biomass remains eligible for Single Farm Payment (SFP) provided that the normal conditions are observed.

If the ECS is included in the new programme, it will be worth noting two constraints that applied under the original scheme:

- The payments were made once only at the time of planting and did not therefore provide income for the ensuing unproductive years.
- Growers had to have an outlet for the produce either with a local power station or in heating equipment on their own properties.

Once SRC has been established, growers on non-setaside land are entitled to an annual Energy Crops Aid Payment amounting initially to €45 per hectare. Different arrangements apply overall to areas covered by Objective 1, which were defined as Cornwall, Merseyside and South Yorkshire. ECS grants, which were available only on areas of not less than 3ha, are paid once the crop has been planted and the agreement with Defra then runs for five years. Notwithstanding the benefit of these grants if reintroduced, a new venture involving the planting of short rotation coppice will need to show a proper investment return in time, including an allowance for the initial start-up period.

Grants are also available for the development of power plants, primarily under the Bio-Energy Capital Grants Scheme, within the BIS. However, there appear to be several difficulties with this, one of which relates particularly to the present context, namely a requirement that energy crops must comprise not less than 25% of total input into a new power station within three years of its construction. Since most woody coppice takes four years to mature, this suggests that developers will have had to be in a position to sign up growers at least one year before starting production.

Another factor has been an emphasis within the grant scheme that developers should be introducing innovative technology rather than use existing systems that are already operating elsewhere in Europe. This inevitably raises the degree of uncertainty and risk of the venture and makes it more difficult to attract commercial funding.

With uncertainty over grant aid, it may be helpful to consider other indicators of the Government's policy on biomass. The Energy White Paper of 2003 refers to biomass as being potentially one of the largest elements in the mix of generation of renewable power, although the original targets for this have since been modified in the Climate Change Programme published in March 2006. This Programme refers to the potential for reducing carbon emissions through the use of biomass for energy and proposes a number of tax incentives on capital costs of CHP plants as well as an exemption from CCL. It also makes a commitment that government buildings should use more renewable energy, some of which will presumably come from biomass heating. In addition, it mentions that a high level seminar on producing methane from waste should be held by the end of 2006.

The topic of biomass was deemed to be of sufficient importance to warrant commissioning a Biomass Task Force in 2004. The report on this, published in the following year, was supportive of the concept of growing energy crops and made a number of proposals for its use, such as installing CHP plants in government buildings and making that also a requirement in the granting of planning permission for new developments. However, it failed to support the concept of using agricultural slurries and other waste to generate methane and other gases. It also decided against the concept of a RO for biomass on the grounds that it would be unworkable due to the fragmented nature of the potential market. Although the Government appears to have an overall commitment towards using biomass as a means of combating climate change, there is little to encourage further private investment unless supported financially through the Development Programme.

Timing

Location is one factor determining the viability of growing biomass, another is timing, in terms of both seasonal production and the length of contract and commitment required from the grower. Some energy crops can be grown within an annual agricultural rotation, notably cereal straw, which is effectively a by-product of a normal arable system. That derives of course from the harvest that has to take place during a specific time, according to the weather and the condition of the crop, with the straw being baled immediately thereafter. The issue of contracts for cereal straw is dealt with in a later section, but in this context it may be mentioned that straw can be readily easily stored, whether in barns or outside stacks. Indeed, this has been a traditional practice for mixed farms when using straw for livestock bedding throughout the winter. The same storage can be used to meet the requirements of the purchasing power plant that is likely to need batches of straw delivered over regular periods during the year. The method of storage will, however, have to be such as to come within the moisture levels specified in the contract.

Miscanthus may also be cultivated as an annual crop although its deep-rooting habit could cause problems at the end of the year if the land needed then to be reinstated for other crops. There could also be time restraints on reusing the land for food production if advantage had been taken of the opportunity of applying municipal waste as a fertiliser on the biomass. It is probably best grown on a perennial basis. Harvest takes place over a wider time period than cereals, between January and April, when the canes have shed their leaves and before they start to send up new shoots. When used for fuel, the crop is also stored in large bales and can be delivered to the power station when required. Harvesting in the winter months has the advantage that there are fewer demands for other arable farm work at the time, so the necessary resources may be more readily available. Operating heavy machinery over arable land in winter can be difficult, given the wet and muddy conditions, as mentioned below in connection with SRC. These problems are mitigated to some extent with miscanthus thanks to its rhizome or root structure, which provides a form of matting just beneath the surface of the soil.

The woody varieties of SRC, notably willow, involve a longer commitment in that they take four years before they can first be harvested and can then remain in production for up to as much as 30 years, although a time span of between 15 and 20 years may be a more realistic commercial estimate. Contracts with processors will therefore need to provide growers with a degree of security over at least a reasonable period of time, especially when bearing in mind the investment that will have been made in coppice plants and possibly also in machinery. The timing of annual production is similar to that for miscanthus in that it is best harvested in the winter, when the stems are bare of leaves and easier to cut. However, not having the same rhizome structure, there is a greater danger of machinery becoming bogged down or damaging the ground under wet, seasonal conditions.

In other countries where SRC is grown, such as in Scandinavia, the ground tends to be frozen in mid-winter, so that any harvesting done at that time should cause only limited harm. In the UK, it is rare to experience any prolonged periods of frosts, especially in lowland arable areas. Furthermore, willow, which is the most common form of woody coppice, tends to grow best on heavier, wetter land rather than lighter free-draining soils, which is better able to withstand the process of harvesting in winter. When cut in winter, the dormant material will be dry enough for almost immediate use. However, in the UK it may be necessary to harvest in the summer as well, and the crop will need to be stored to allow it to dry out sufficiently before it can be processed. This will require both time and space, and may make the process less effective.

The question of timing of such crucial operations needs to be assessed not only in terms of being able to supply the processing plant in a regular and economic manner, but also as to whether it introduces further potential problems. Summer harvesting can lead to possible contamination of the fuel, especially through mould forming during storage, as well as a decline in fertility due to the regular removal of leaves and other green material that would otherwise form a mulch or compost over the root stocks.

The crop may be stored on the farm in the open as cut stems or "billets" before being processed and delivered to the power plant. Harvested coppice is likely to be fired in the form of chips or pellets rather than as cut stems or "sticks". The crop can be harvested as sticks and then chopped at a later stage, although it would be more usual for it to be chopped at harvest and then delivered in the form of chips. Essentially, the cut coppice has to be stored after it has been harvested in order for it to dry to a suitably reduced moisture level for efficient burning and this process can then also improve its transportability. It is mostly stored as loose chippings, in the open or also in farm buildings where available.

Timing can also be a consideration when sourcing biomass from woods, especially if the purchasing contract requires a regular supply of chips throughout the year. Forestry operations tend to be less regular than in farming and generally less well resourced in terms of labour and machinery. These issues are mentioned further in the section of forestry (towards the end of the chapter).

Many boilers and power plants will be operating on fuel in the form of pellets rather than chips, which necessitate an intermediary process, as mentioned later.

Commercial production opportunities

For the farmer, the opportunity to cultivate biomass as an energy crop will initially depend on whether a suitable power plant exists within the area or is due to be opened there, for reasons both of economy and for securing grant aid, as mentioned in the previous section. The question of economics is likely to be true for the woodland owner too if thinnings, traditional coppice and other small material are to be chipped on site. The considerable cost of hauling felled timber to a distant sawmill may be warranted when that timber is converted into planking and other finished products. For wood chips, however, the end value will be far less even, though the haulage costs would be broadly the same. It is unlikely to be feasible to produce and deliver chips for fuel other than over relatively short distances.

Biomass is best used to produce power and heat on a local scale in purpose-built plants, such as is already being done in a number of countries in northern Europe. In the UK, this has got off to a slow and uncertain start, suffering a major setback when ARBRE, the company developing the first major wood-burning plant near Selby in Yorkshire, went into liquidation in 2003 after only eight days' production. A number of reasons have been cited, including two aspects of government policy:

- A requirement to use new technology in preference to buying adopting systems from abroad.
- The fact that the RO focuses almost exclusively on the sale of electricity.

The emphasis on electricity, which has been mentioned also in chapters 1 and 2, is largely due to it being relatively easier, and cheaper, to regulate than other forms of power. The new technology such as that being developed by ARBRE, was said to need more funding than the DTI appeared ready to give. There was difficulty too in adhering to the preferred policy of generating electricity in that, as stated earlier, biomass is better suited to producing heat rather than driving turbines for electricity. Since then, however, other initiatives have been taken and it is hoped then that more such plants will be developed in the coming years and that the opportunities for growing biomass will increase accordingly.

At present, some biomass in the UK is being delivered to existing large-scale, coal-fired power stations for a process of co-firing where coal is blended with a combustible organic material such as sawdust. Although this then reduces the carbon dioxide emissions of the station from the level produced by coal alone, there are a number of limitations on this system, which suggest that co-firing may be of only limited significance to most individual timber or biomass growers.

Markets

The planting of SRC in the form of willow or poplar will, at present, be viable only if a secure contract can be obtained from a power station within a close radius of the land or if a smaller heating plant is being established for more localised use, possibly on the property itself. Miscanthus, on the other hand, does have other uses than just fuel and can be grown for equine and other animal bedding. As such, it may be feasible to consider planting miscanthus without a secured outlet to a power station, but perhaps in anticipation of one being constructed. It could also in the meantime be contracted out to a power station that is planning to adopt co-firing.

Miscanthus can also be treated as an annual crop, as mentioned previously, and may be cultivated with existing farm machinery, thereby involving a lesser investment commitment than willow coppice, which has to be established over a longer term and harvested using specialised machinery. However, a crop of miscanthus can produce commercial yields for up to about 20 years. The minimum term of agreements under the ECS has been five years.

A new power plant is currently being built at Eccleshall in Staffordshire, which is set to operate mainly on miscanthus that is to be grown by a cooperative of farmers in the locality. It is expected to use 22,000 tonnes of miscanthus per annum and produce around 16,000MWh of electricity. Most such plants are designed to work with a diversity of fuels, including wood chips or pellets and miscanthus and also some organic wastes. They may be either as a single commercial unit providing CHP or heat alone or as "clusters" of boilers producing heat for a group of premises, such as council flats and offices. Some of the latter plants are now in operation, but other larger schemes are still going through the lengthy planning stage, such as a 23MW power station using gasification at Winkleigh in Devon and another near Swindon that proposes to generate 2.5MW of electricity and to utilise the surplus heat to dry wood fuel for other plants.

Processing

The basic process of firing biomass is through ordinary combustion, when the heat derived is used to produce steam, which then drives a turbine to produce electricity. This system is fairly straightforward and well established, but also it is relatively inefficient in converting woody material into electricity. An advance on this is gasification, in which the burning is controlled by restricting the supply of air so that combustible gases are given off and used to power the turbine. Another development is pyrolysis, where the wood is heated without oxygen so that it degrades into a liquid, which can be used as an efficient heating agent together with by-products of gases and charcoal.

These two latter technologies are able to improve the efficiency of conversion of wood into electricity, but they are still at an early stage of development and not in full commercial production. Under current methods, the output from one tonne of wet wood, which can produce about 2.5MWh of heat energy, would be reduced by around two-thirds if it were converted solely to electricity.

Co-firing

Operators who adapt conventional power stations to co-firing are able to offer ROCs for a proportion of the electricity that is thereby generated, and their customers may benefit also from gaining some exemption from CCL. Therefore, there is a financial incentive in adopting this process. Co-firing involves the use of a blend of coal and biomass as fuel, which produces lesser carbon dioxide emissions than when coal is burnt on its own. Existing power stations can function on fuels that include a blend of up to about 5% of biomass material with only minimal modification, but would need more extensive redevelopment if a higher proportion of wood were to be used. Bearing in mind the age and limited economic life of most coal-fired stations, it would seem unlikely that the additional investment needed to take on larger quantities of wood could ever be justified, even with a potentially rising premium from ROCs. The intention of current policy would seem to be to use these existing power stations as a means for initiating the establishment of SRC as well as the use of other timber by-products. Indeed, under current regulations, co-firing will be eligible for ROCs only until 2016. Then, as the existing stations are decommissioned, the coppice and other materials could create a supply source for CHP plants that may have been purpose built in the future. Meanwhile, a considerable area of SRC would be needed to service even that limited blend of up to about 5%.

One example of a conversion to co-firing is at Didcot in Oxfordshire, where a supply of over 30,000 tonnes of energy crops is expected to be needed each year. If this were to be grown as SRC, an area of at least 3000ha, or about 7,500 acres of land would have to be planted, within the preferred radius of around 25 miles. In Yorkshire, around 2000ha were planted to supply the ill-fated ARBRE plant and may now be destined instead to be co-fired in the Drax power station near York. Originally, the ECS was funded in 2000 to enable 20,000ha of coppice to be planted across the country. These rough figures assume that the co-firing will use only locally grown material, but also viable alternatives are being imported in the form of wood pellets from Sweden or palm nut kernels from the Far East and even olive stones from southern Europe.

Co-firing was introduced as part of the RO of 2002 and then reviewed in 2004 when it became clear that there was a shortage in supply of energy crops. Under the modified requirements, power stations can qualify for ROCs if they incorporate a 5% mix of biomass into the coal. Of this, a growing proportion needs to come from dedicated energy crops, as opposed to other materials as mentioned above. These requirements, beginning at 25% in 2009 and rising to 75% in 2011, have been brought under review again in 2006, due to concerns that the large coal power stations could buy up a disproportionate quantity of the limited stock of available biomass and put other processors, such as the fledgling CHP plants, at a disadvantage. This, together with other measures that effectively cap the amount that can be claimed under the RO, raises some uncertainty over the viability of planting SRC specifically for co-firing. Interestingly, the majority of applications under the ECS have been for miscanthus rather than SRC, suggesting that there has been a preference for growing a crop that requires a lesser commitment in time and in acquiring specialist equipment.

Cultivation

Woody coppice

Woody coppice that is being grown as an energy crop tends to be willow, which is easier to cultivate for this purpose than poplar. The latter might be used on particular sites where it may be better suited or to provide a visual mix within a plantation that is predominantly willow. It needs to be established on cultivatable land that can be prepared as if for an arable crop.

Willow is relatively tolerant of soil type, ranging between 5.5 and 7pH, but it does require moisture and is likely to do best in areas where annual rainfall averages 600-1000mm. The young plants, which are in the form of nursery reared "rods", are vulnerable to rabbit damage and fencing may therefore be needed. Weeding is also necessary initially, and the new plants are cut to ground level at the end of the first year to encourage tillering. Fertiliser is likely to be applied at the outset and then after each three yearly harvest.

Fertiliser

Being a non-food crop, animal slurries and municipal waste, such as sewage sludge, may be used for this purpose provided that local water conditions permit the use of materials of this kind. Indeed, the disposal of treated sewage is becoming a growing issue among water companies and public authorities and a solution such as this could become a driving factor in the cultivation of SRC for woodfuel. It may be easier to use these materials on biomass than on food crops, but there are still concerns that small traces of heavy metals and other toxins could be released into the atmosphere when the crop is being fired. Modern burning processes, however, that use gasification tend to eradicate even these particular elements and so make it more feasible to use coppice that has been fertilised with sewage sludge.

The crop itself can reduce the leaching of nitrates and other matter, thanks to the wide spreading nature of the roots and the long growing season of the plant compared with conventional arable crops. Most of these operations are carried out by specially adapted machinery, probably under contract. Harvesting is ideally carried out any time between October and March. This is after the leaves have fallen and before the new buds emerge, and it comes conveniently after most other arable operations have been completed.

Mostly, the matured coppice is harvested into wood chips, using a process not unlike that used for forage maize, whereby the stems are cut and chopped and blown into a trailer behind the harvester. It can also be cut into rods, which may be easier to handle on the farm, but which can result in a lower quality product, as the dried rods have a tendency to shatter when chipped.

Some heat conversion processes, especially the smaller ones, are more effective and easier to run when fuelled with pellets rather than ordinary chips. Whereas wood chips can be created from a single physical process of grinding down any variety of woody material, pellets are manufactured essentially from sawdust, which is then bound together with natural lignins. The latter have the advantage of being easier to use both in power plants and individual boilers due to the fact that they have a characteristic of flowing like a liquid, whereas chips need more heavy handling. Pellets also have a more consistent quality and may be easier to transport and store, but have the disadvantage that they can be more expensive to produce in terms of money and carbon emissions. Until recently, all wood pellets were imported but a number of production plants have now been established within the UK.

When harvested at the end of each three-year cycle, willow coppice should yield about 30-35 tonnes per hectare, with a current market value of around £30 per tonne. This usually refers to the weight once the crop has been oven dried and the cost of drying together with that of harvesting and chipping amounts to about £10 per tonne.

Miscanthus

The growing of miscanthus is similar to willow coppice, although the initial stock comes in the form of nursery-raised rhizomes rather than "rods". The plant tolerates a range of soil types but is sensitive to drought and to late spring frosts. This latter characteristic means that it is best grown in southern England although it is also being planted in Yorkshire and trialled in Scotland. It requires a little fertiliser on establishment, but thereafter it provides most of its own nutrients through leaf mulch, which also helps to control weeds. The crop, which grows to about 4 metres, is harvested annually during the winter and early spring when the canes are bare of leaves and before the cut stems are ready to produce new shoots. When being harvested for fuel as opposed to other uses, such as animal bedding, the crop is cut by a mower and baled. The large bales can then be stored in the open until consumed.

An established crop of miscanthus is likely to yield between 12 and 15 tonnes of oven-dried material per hectare, which has been priced latterly at just over £25 per tonne for fuel but can sell at a higher figure for equestrian bedding. In milder areas, the crop can be harvested one year after planting, albeit at a reduced yield, although in other areas full production may only be achieved after three or four years.

Straw

Cereal straw is already potentially available on most arable farms and could be sold to a power plant whenever the opportunity might arise. The economics of transporting straw for fuel allow for a slightly wider radius than that for SRC and, for example, the Elean Power Station in Cambridgeshire is able to acquire supplies from a distance of up to 50 miles. Some management issues might need to be considered, such as whether there would be a disadvantage in not incorporating the straw into the soil or using it for cattle bedding and thereby manure, and whether the ongoing arable enterprise would be able to fulfil the terms of a fuel contract, as regards quality and timing.

Straw is often considered as a secondary by-product of cereal cropping that can be disposed of relatively rapidly even under unfavourable conditions, particularly when there is a need to clear the land as quickly as possible before autumn cultivations. Contracts for supplying power plants are likely to include rigorous terms as to the moisture content of the baled straw and the timing of deliveries. There could also be a question of the period of years over which a grower might be required to supply straw to a power plant. Under current conditions, farms that are growing cereals may gain a net advantage in being able to sell straw to a local power station for a price that exceeds the cost of baling and transport, rather than depend upon some other means of disposal.

There is a view that changing circumstances in agricultural policy and world grain markets could lead to cereal production becoming unprofitable for many farmers. If this were to happen, then growers would be advised to cease production and merely keep their land in "good agricultural condition" in order to continue to benefit from support under the CAP. That could then lead to breaking an agreement with the power generators, who would have been depending on receiving a steady supply of straw. Alternatively, the affected growers might let out or contract their land to other local farmers, who could cultivate cereals with the benefit of improved economies of scale but who might be unwilling to take over the commitment to supply straw. It may be unlikely that such a situation will arise on farms that are already operating efficiently on a large commercial scale, but it may need to be considered when undertaking to supply straw for generating heat or power.

Poultry litter

Poultry litter can be used as fuel for generating electricity, although the facilities for doing so are still limited in number. At present, it is therefore an option only for relatively large poultry units that are located within a reasonable distance from one of the few purpose-built plants that can utilise this material, such as in Fife, Suffolk and north east Lincolnshire.

The incentive for developing a means for disposing of this type of litter derives partly from the fact that it may become increasingly difficult to spread poultry manure on land, especially in Nitrate Vulnerable Zones (NVZs). The combustion process produces an ash that is high in phosphates and potash but relatively low in nitrates, and so can be more readily spread onto farmland than the original litter. Further plants are being planned that will be able to operate on a mix of litter alongside other forms of biomass. The commercial poultry market can be volatile and while the sale of litter may seem to be a useful added value to the existing business under prevailing conditions, it could become more difficult to fulfil if economic conditions were to change.

Anaerobic digestion

Producing methane gas through the controlled decomposition of animal and food wastes may not be thought of directly as a form of biomass or energy crop, but it can fulfil a similar function in agriculture and is therefore best considered in this present context. The process of anaerobic digestion through which this is done produces a mixture of methane and carbon dioxide, in a ratio of approximately 2:1. The carbon dioxide then has to be removed, leaving what is known as "substitute natural gas", which can be harnessed to generate heat and thereby electricity, albeit with only about half the effectiveness of a mineral-based, or "natural", gas.

Although the economics of anaerobic digestion have been discouraging, the scope for creating biogas in this form may now be improved by a recent European Directive on Animal By-products, which has broadened the categories of organic wastes that will need to be treated by specific processes that include the production of gas.

Wet biomass

Wet biomass refers primarily to animal slurries and to food waste that can be converted into biogas by means of anaerobic digestion. While the technology for this is being employed in other countries, notably in Germany, it has only recently been introduced to the UK with the development of a pioneering anaerobic digestion plant at Holsworthy in Devon. This is designed to operate off a mix of cattle, pig and poultry manures, as well as abattoir and organic food wastes, which are sourced within a relatively short radius of no more than six miles. The residues that are deposited after processing are useable as agricultural fertilisers and are returned to the farms. This cycle can benefit the producers by allowing the slurries to be stored and spread over a more convenient time-scale than if they were being used directly on the farm. The plant was built originally as a community project but failed to sustain the necessary funding and is now being revived under a commercial ownership.

A similar plant is in operation in Ayrshire in Scotland, where the output from a group of local farms in being processed successfully into methane and used for domestic heating and generating electricity. The finance for this was provided in total by the Scottish Executive largely as a means of preventing any agricultural effluents from polluting the local beaches.

The economics of investing in these digesters and in the heating infrastructure, without substantial grant aid, still seem to be commercially difficult in the present market. However, a pioneering plant has recently been developed on a farm near Bedford that produces electricity and bio-fertiliser from pig slurry and organic waste. The developers in this case reckon that it needs to be supplied by not less than about 300 cows or 5000 pigs and have an area of around 500ha upon which to spread the bio-fertiliser. It is also possible to install individual anaerobic digesters on site and to produce gas for consumption around the farm.

Landfill gas

Gas can also be harnessed from municipal waste that has been deposited in sites such as former quarries and then been covered and prepared for reinstatement into farmland. The degradeable organic material decomposes to produce a gas that contains a mix of methane and carbon dioxide, which can be extracted and used to generate heat or electricity. However, this tends to be viable only in certain situations, on larger sites incorporating an appropriate type of waste. Where there is potential for drawing off gas on a commercial scale, it tends to be developed by a major power supplier working together with the operator of the quarry. The landowner's interest in this may need to have been defined within the original quarrying agreement.

While methane is a useful by-product of waste that has to be disposed of, the process of anaerobic digestion produces almost as much carbon dioxide as methane. However, the potential greenhouse effect of methane is said to be more than 20 times that of carbon dioxide and there is likely therefore to be increasing pressure to find better means of controlling gas escaping from landfill sites other than by burning it off at the surface. Using such gas to fuel CHP plants within the local community may be restricted by the fact that landfill tends to take place on sites that are removed from residential areas.

Biomass from forestry

Government policy on biomass tends to focus on the agricultural sector, with grant aid being offered for the growing and processing of SRC and farm waste, rather than on forestry products. Such support as there is comes more in the form of helping to create an end market for wood chips and other timber materials, by offering grant aid and tax breaks towards the installation of wood boilers. These are referred to in further detail in the section on small-scale production.

Mention has already been made of other measures that should encourage the use of woodfuel, such as co-firing in coal power stations and the intention to require public buildings and new developments to use renewable energy systems. However, despite the lack of much direct Government support, there is considerable potential in this area.

Residues

Woodland management is essentially about producing timber, although it does now also encompasses an increasing number of other facets, such as amenity, conservation and shooting.

Conventionally, when a stand of timber reaches its commercial maturity it is felled and sold and the ground upon which it was growing is then replanted or reseeded. In the years before felling, the plantation might be thinned in order to enable a selected final crop to develop to its full potential. This thinning process depended upon there being a market for the semi-mature trees, which could be processed into products appropriate to their size, such as fencing bars or pulp. Any surplus material that had been trimmed from the main trunk would previously have been burnt on site.

Latterly, much of this branchwood tends to be left within the wood in order to save on labour costs, to reduce the emissions from burning as well as the risk of a fire spreading, and to provide habitats for invertebrates and fungi. (While this practice is seen to be a positive step towards environmental improvement, there is a counter aspect to it in that when the trimmings and other remaining material decompose, they give off methane, which is one of the most pernicious greenhouses gases.)

This secondary material may be cleared from the site by being ground up or chipped into a product that has uses in the manufacture of composite, or "chip", board, as a horticultural mulch or horse bedding or as fuel. While such residues may appear to be an inexpensive by-product of commercial forestry, especially compared with the dedicated cultivation of coppice, its collection is relatively labour intensive and if the cut branches are twisted or contain knots they consume more energy when being ground into chips. Wood from either broadleaves or conifers produce an equivalent amount of heat per tonne when processed, although the conifers are of a less dense material and will therefore involve a greater volume of material. Waste, or "arisings", from arboricultural work such as tree surgery and park maintenance can also be used as a source of biomass fuel.

Woodfuel

Woodfuel comprises a number of forms, ranging from traditional logs to compacted pellets. Some of the more refined forms can be used to generate heat on a commercial scale, as in the case of coppiced biomass mentioned in the previous section. The potential for using forest products in this way faces similar constraints to those for SRC, notably the distance it has to be transported and also of the condition and quality of the material to be used as fuel. There may be an additional difficulty in the case of forest residues, as it can be harder to deliver a constant and regular supply from timber operations than from a rotated "arable" crop such as SRC.

In all but the largest holdings, felling or thinning operations tend to be carried out on a sporadic rather than annual basis, determined partly by the specific trees being grown within a wood at any period in time and also by the vagaries of the timber market. If prices for a particular type of timber are low at the time when the trees might be ready to be felled, then the owner has the option to leave them standing for a further season or more in the hope that better prices could be achieved in the future.

Although the volumes that are then produced during these operations may be substantial, the chips would need to be stored in order to provide a reservoir of constant supply for the heating plant. Such storage requires space as well as ongoing management in order to avoid loss or deterioration. Furthermore, even if such a method of production were to be used, the supplier would be committed to replenishing the store on a regular basis and thereby perhaps be forced into having to fell or thin crops at a time when it would be uneconomic to do so.

Timber that is felled and sold to a sawmill for processing may also produce material for fuel, albeit at one later stage in the system. The offcuts arising from the debarking and trimming of sawlogs can be converted into chips and the sawdust itself can be used for blending with coal. These residues are a considerable resource in that they amount to almost half of the wood that leaves a sawmill. Sawdust is effectively a natural by-product of any timber that is cut into a useable form and would seem to be an inexpensive and constantly available source of biomass. However, it does have one drawback in that it absorbs moisture more readily than chips or pellets, making it less efficient as a producer of heat. That loss of efficiency can be mitigated by storing the sawdust under cover, but providing such storage requires an additional capital investment, effectively raising the cost of sawdust and reducing its initial competitive advantage.

In order to be used as a commercial fuel, sawdust needs to be manufactured into pellets or blended with coal, which relies on the fact that co-firing in power stations is eligible for ROCs. That eligibility has been granted only until 2016 and requires that such fuels are blended on site at the power station. This means that generators will not only have to make the necessary investment in covered storage themselves, but also will have to recoup that expenditure in a relatively limited time frame, which may be unfeasible.

Coppice

Growing coppice within woodland has been a traditional practice for many centuries, although latterly it has been largely abandoned due to a lack of sufficient outlets for its products. Wooden wheel spokes, fencing spars and domestic wood fuel are no longer required in any commercial capacity and growing coppice for these purposes has dwindled to an insignificant level. The species mix associated with traditional coppice, such as hazel growing under oak, is still present in many broadleaf woods and efforts are being made in some areas of the country to bring the coppice back in to production.

This tends to be a labour intensive, and therefore costly, process and it is difficult to see how it might achieve a significant recovery while the prices for its products, such as charcoal or fencing materials, face severe competition from imports and substitutes. Some financial support is available via grants from certain local authorities and by the Forestry Commission, although the latter are now restricted. Growing traditional coppice as a fuel for CHP plants therefore depends heavily upon being able to command a viable price within the power industry, as well as being offered a secure contract from a processor located within a viable distance from the woods.

Timber waste

Landowners and foresters are concerned essentially with the primary source of timber, namely the growing and harvesting of trees, but their position may also be linked to the opposite end of the cycle, where timber products are being removed and destroyed.

When buildings are being demolished or furniture is being replaced, the timber will, in the recent past, have been disposed of in landfill sites. The pressures on landfill throughout the country have given rise to a tax on material being dumped there and this has encouraged industry and public authorities to consider alternatives. As the tax is set to rise steadily over the next few years, from £15 per tonne to ultimately £35 per tonne, the incentive to find alternatives is likely to increase accordingly too. In the case of timber products, this will involve recycling, generally in the form of woodchip or pulp. The need to find outlets for such residues should give a further impetus to the development of CHP plants and to the markets for timber by-products.

Small-scale production

There are many appliances available on a domestic or small commercial scale that can operate on fuels comprising timber or straw. These are essentially boilers that provide heated water rather than electricity, which is a function that is only feasible on a larger industrial scale.

Having an individual heating unit should qualify a grower for grant aid for the establishment of SRC. The installation of the boiler, whether for domestic or community or business use, may also be eligible for grant through the Energy Saving Trust (under the Low Carbon Buildings Programme in England or the Wood Energy Business Scheme in Wales or the Scottish Communities and Householder Renewables Initiative).

The amount of funding being offered varies between the national schemes and according to whether it is for domestic or community use, but range from £600 or 20% for a simple installation in England to a maximum of £10,000 for a larger project in Scotland. Businesses that invest in biomass heating can benefit from an enhanced capital allowance whereby 100% of the cost may be offset against taxable profits in the first year.

There are opportunities too for incorporating biomass into a farm diversification project, such as for heating a visitor centre or tourist accommodation. This would not only make it a more sustainable development, but also increase the chance of securing planning permission as well as possible funding under the renewed Development Programme. The volume of woodfuel that might be required will depend upon the scale and design of the individual project, but may range from about 100 tonnes per annum for a large house and outbuildings to perhaps 10 times that amount for larger venture such as a rural educational facility. A boiler that could provide for the heat and energy needs of an individual farm by using miscanthus, would require a cropping area of approximately 20ha. There are heating systems on this scale that can be fuelled with straw and now also with the cereal grain itself, notably wheat. The latter are being developed in continental Europe and offer the advantage in a domestic context that the fuel needs less storage space and is easier to handle than straw or woodfuel, although creating a greater amount of ash and clinker for disposal.

Action points for farmers and landowners

- Contact local producer groups or potential buyers.
- Check the suitability of land for growing SRC.
- Check the availability of grant aid.
- Consider the effect on existing farming operations of committing land to longer-term contracts for SRC, straw or animal litter.
- Assess the anticipated costs and returns, including the receipt of grant aid.

Action points for woodland owners

- Contact the potential buyer.
- Assess the implications of being required to supply material regularly and under longer-term contract.
- Consider arrangements for the provision of machinery and labour.
- Check the availability of grants for managing woodland coppice.

basis, then they will effectively absorb those carbon emissions and the process can thereby be deemed to be carbon neutral. Since road transport is estimated currently to be producing as much as 22% of all greenhouse gases across Britain, the use of biofuels could make a significant contribution towards meeting environmental targets.

The UK market

Arable crops can be converted into both diesel (as biodiesel) and a petrol substitute (bioethanol or biobutanol). These are generally blended with ordinary mineral oils, although it is possible to run existing vehicles totally on biodiesel. In the US, car engines are being modified so as to operate on a mix containing up to 85% bioethanol, made principally from maize. In Brazil it is mandatory for all petrol to contain at least 25% bioethanol, which is derived in this case from sugar cane and which now accounts for around 40% of total consumption, with some cars being designed to run on 100% if required.

In Europe however, the level of mix tends still to be at only 5%, due to limitations imposed in the warranties from engine manufacturers, which are also incorporated within EU quality standards. However, in France, for example, just over half of the fuel sold on forecourts comprises a 5% blend, which in total must then add up to a significant reduction in carbon dioxide emissions overall. In France too bioethanol is being used in an etherised form as a petrol extender. Growing a totally renewable source of low emissions fuel on existing farmland, using existing agricultural equipment and systems of cultivation at a time when farming profits are under pressure seems to be a perfect solution. But it is only recently that it has started to be developed in the UK.

Economics

The main reason for this apparent delay is cost. The expense of the initial investment in developing new refineries to process crops that have then to be bought in at the equivalent of prevailing food market prices means that the biofuels so produced would have to retail at a higher price than existing mineral oils. As much as half of the consumer cost of road fuels is, however, made up of tax in the form of excise duty. In those other countries where biofuels are being used commercially, this tax has been reduced as part of a national environmental policy.

In the UK, this has been resisted, due perhaps in part to fiscal priorities and to the fact that oil and gas is being produced "locally" from the North Sea. However, in his 2004 Budget the Chancellor of the Exchequer announced a reduction of 20p per litre in the level of duty on biofuels off the full amount of 47.1p being levied on mineral oils.

This apparent breakthrough in enabling the commercial production of biofuels was, however, tempered by two immediate drawbacks:

- The measure was not going to be introduced until the beginning of the following year.
- The amount of reduction was too small.

It was certainly less than the 40p concession allowed on gas fuels such as LPG and it was also, more significantly, below the figures of 30-35p that the industry estimated as being needed to enable biofuels derived from arable crops to compete with mineral oils. It seemed initially that the 20p concession was likely to benefit only the production of biodiesel from RVOs, which has lower processing costs than working from agricultural crops, and which has subsequently become ineligible for the concession. The Budget measure did little therefore to progress the manufacture and sale of biofuels, even though it appeared to demonstrate Government support.

The Renewable Transport Fuel Obligation

There has been a further political development that has brought a new potential impetus to the matter. The Energy Act 2004 included a facility whereby the Government would have the power to introduce a Renewable Transport Fuel Obligation (RTFO). This follows a European Directive requiring member states to incorporate certain biofuel measures into national legislation before the end of that year. When implemented this new Obligation would, as in the case that already operates for electricity, require suppliers of transport fuels to include within their annual sales a specified percentage that was derived from renewable sources. The outcome will be, as with electricity, a guaranteed market each year for biofuels whose production costs could be subsidised by the value of the ROCs that were secured in their manufacture.

Oil companies will need to obtain certificates from a Government agency to demonstrate that a certain percentage of fuel sold each year was derived from renewable sources. If their sales exceed that percentage, then they would be able to sell the surplus certificates to other companies who were unable to fulfil the target amount. To qualify for these certificates, it will be necessary to produce evidence of the level of carbon savings achieved and of the sustainability of the biofuel. This may give some protection for home-produced fuels against the threat of cheaper alternatives, such as palm nut oil being imported over long distances from less sustainable sources.

Commercial output should then become feasible, especially during a period of relatively high prices for mineral oils, as this then reduces the competitiveness of conventional petrol and diesel. On the other hand, if pump prices are high it becomes more difficult to raise them further by the amount of the cost of the ROCs and, as in the case of electricity, the viability of the RTFO could come into question. In Britain, the RTFO is due to be introduced in April 2008 at a level of 2.5% rising to 5% by 2010. That could then increase further according to proposals made in the Government's *Energy Review* of 2006 and it is estimated that if this were to reach 10% by 2015 it would save around one million tonnes of carbon, which is the equivalent of taking a million cars of the roads.

The introduction of the excise concession and the RTFO have, together with the current state of the oil market, encouraged the development of a number of biofuel processing plants within the UK. These are due to begin production in 2007 but they have, prior to this, already been offering contracts to growers on a forward basis. Some of these cover not only the initial year, but also include incentives to commit supplies for the following two years. These contracts are essentially for oilseed rape, wheat and sugar beet, as itemised later. The length of term, such as over a three-year period, may be determined by the need for developers to be able at the outset to demonstrate for development funding purposes that they have a guaranteed ongoing business. It may be that contracts will in future tend to be for single seasons and in greater competition with imports of oilseed, ethanol and other oils from abroad.

Meanwhile, demand has already been growing as public authorities and institutions in particular are looking to use "green" energy on the roads as well as within their buildings. The stage is finally set for a British biofuel industry that will, according to the British Association for Bio Fuels and Oils (BABFO), need around two million tonnes of crops if it is to fulfil the 5% requirement. There is a questionmark over how much of this may be sourced nationally as opposed to imported from potentially cheaper sources abroad. It may be no coincidence that a number of the new plants are being developed close to ports, such as on Merseyside, Grangemouth and Humberside.

BABFO has suggested that the current estimated requirements could be achieved with home-grown feedstocks. Some one million tonnes of biodiesel could be produced from oilseed rape on land that is currently under setaside and another one million tonnes of bioethanol refined from the three million tonnes of wheat that are exported in most years from the UK. That may be an indication perhaps of how feasible it would be to meet the current targets, even if in practice using all the setaside land in the country for growing industrial rape might not be considered particularly attractive. There should be little problem either in sourcing sugar beet for processing into fuel, bearing in mind the ongoing international debate about reducing tariffs and prices for sugar beet grown in Europe so as to enable sugar cane to be more freely traded from the developing world. It is likely now that a range of relevant agricultural crops may be grown on ordinary farmland as well as setaside and then sold to the fuel market in parallel with foodstuffs.

Combineable crops grown for biofuel are currently eligible for an annual Energy Crops Aid Payment (ECAP), which was introduced within the EU in 2004 and currently amounts to €45 per hectare. This would equate to around £6 per tonne and be payable on any arable area that has been planted with an eligible crop and contracted to a fuel processor. It provides a guaranteed premium, and incentive, for the grower over and above the normal food market price. It is not available for sugar beet or for crops grown on setaside land. If crops are grown for fuel on setaside land and then fail to be sold for such use they cannot be diverted into the food market.

ECAP is available specifically for the cultivation of annual arable crops for energy uses, which may include heating and biogas as well as biofuels. Payment may be claimed upon evidence of the crop being contracted to a processor, and may be received either by the grower direct or within the terms of that contract. ECAP is separate part of the Energy Crop Scheme (ECS), which applies to perennial crops such as short rotation coppice and miscanthus, as outlined in chapter 4.

It would be technically possible to produce liquid fuel from a solid biomass, such as miscanthus or woody coppice. However, there is little indication at present that this could become as commercially viable as using arable crops and it is more likely that in the foreseeable future these materials will be used instead for generating heat and power, as discussed in chapter 4.

Contracts for supplying crops for processing into biofuels will need to include evidence that they were grown in a sustainable manner. The conditions for this are likely to be similar to those for Crop Assurance Schemes or Linking Environment and Farming (LEAF), in terms of having incorporated good agricultural and environmental practice, but also may include a form of carbon certification. This will mean providing a record of the inputs that went into growing the crop, notably fuel and fertiliser, as part of an overall assessment that there has been a net saving in greenhouse gas emissions by the time the end product is consumed. That could mean the farmer having to adapt to certain specified farming practices regarding dessication of oilseed rape or using minimal cultivations for example.

Biodiesel

The production of biodiesel from agricultural crops appears to be remarkably straightforward, in that it can be done with conventional oilseed rape using an established processing system. Any Double Low variety of rape would be suitable, although the market is unlikely to accept genetically modified crops, since those end users who are inclined to buy a "green" fuel may also have views about the principle of genetic modification.

By the same token, growers of rape for biodiesel are likely to be required to provide certification about the level of emissions emanating from cultivations. This would be a record of the fuel and fertiliser used in establishing and harvesting the crop so that buyers of the end-product could be reassured as to the level of net carbon savings gained by using the fuel. Such certification may be needed in order to satisfy the purchasers of the diesel that they are buying a sufficiently "green" product and also to qualify for the RTFO.

High Erucic Acid varieties are technically unsuited to conversion into diesel, but rapeseed oil otherwise produces the highest quality diesel in preference to, for example, that derived from soya or from RVO. The seed is crushed into oil and then processed directly into biodiesel in the form of Rapeseed Methyl Ester (RME). In principle, farmers should be able to offer their rapeseed for sale to diesel processors, including both that which was grown on setaside land as an "industrial" crop and the normal "commercial" output. In practice, however, there is the question of price.

Past experience has shown that the cost of producing RME has been greater than that of refining mineral oil. Now the balance seems to have changed due to a combination of factors:

- The price of oil.
- The availability of tax concessions and grant aid.
- The introduction of the RTFO.

A number of biodiesel plants are being developed within the UK that may use a mix of feedstocks but will certainly be requiring supplies of locally grown oilseed rape. Also, it has become more viable to install on-farm units for converting rape into diesel.

Commercial markets

Oilseed rape crops may be sold to fuel processors in much the same way as into the conventional oil and foodstuffs markets. There could be some limitations according to location, as there are still only a few biodiesel plants in the country and unless suppliers are within a reasonable distance of one of these they could face higher transport costs than when selling into the more general market. Most varieties are equally well suited to processing into fuel, other than those high in Eruric Acid or that may be Genetically Modified.

Price will always be an issue, as the crop will be competing not only with conventional agricultural markets, but also will be dependent on the price of oil. As an indication of this, it was estimated before the introduction of the RTFO that when mineral oil was trading at between $40 and $50 per barrel, the price of rape would need to fall to about £110 per tonne, or by over 20% from its prevailing level of around £140, before it could be used for commercial production of biodiesel in the UK. There will be ongoing competition too from imported vegetable oils, whether from rape or soya or tropical products such as palm nut oil.

On-farm production

It has now become feasible for biodiesel to be produced on a farm scale, as only a reduced rate of excise duty would be levied on it. Equipment and expertise are both available to crush and distill rapeseed onsite in quantities upwards of one tonne and producing between 300–2,500 litres of biodiesel. The cost of doing so currently works out at around 60% of "pump prices" of ordinary diesel, after including just the concessionary level of duty at 27.1p per litre. However, this will be more than the price of "red" diesel, which is still being sold commercially for agricultural use at a greatly reduced rate of duty that does not seem to be available to individual producers.

The initial capital outlay ranges from £15,000-40,000. There are possibilities for one such plant to produce diesel to meet the needs of a group of farmers or a local filling station. The process also creates by-products that will need to be disposed of, notably glycerine and rapemeal. These may need to form part of the financial return if the whole exercise is to be viable and there could be difficulties in finding suitable outlets from a rural location, especially as the use of rapemeal in cattle feed is relatively limited. ECAP is available for crops processed on-farm, subject to certain controls such as weighbridge measurements and separate accounting.

Biodiesel has the advantage that in most engines it can be used totally as a substitute for conventional diesel, so locally processed fuel can be filled directly into farm and other vehicles. Bioethanol, on the other hand, tends to be used only as a minor additive to petrol.

Bioethanol

The potential in the rural environment for bioethanol should be even greater than that for diesel in that it can be made from more than one type of arable crops. In the UK, this focuses on wheat and sugar beet, although under the right conditions other crops could also be used, such as potatoes, maize, barley and rye. Wheat is already being distilled into bioethanol in Britain, but this has been almost exclusively for the drinks industry since, under the prevailing fiscal and market conditions, its cost has been above that of petrol. There are also some useful by-products being derived during processing, such as animal feed and industrial carbon dioxide.

As in the case of biodiesel, the introduction of the RTFO and concessions in excise duty together with the rise in oil prices have brought about a radical change in market potential. Processing plants for converting wheat and sugar beet into ethanol for use in petrol are under construction, with contracts for crop supplies being offered from 2007. As with biodiesel, the crops that can be used for bioethanol may be sourced from ordinary agricultural cultivation, whether on setaside land or from mainstream cropping. Any variety of soft wheat can be used for this process, although those that are high in starch give the highest ethanol yields. Indeed, some of the pioneering contracts for feedstocks for bioethanol are offering a premium of around £3.50 per tonne.

As this market develops, more information is being gathered about which crop varieties are best suited for the purpose and about the agronomy to be used in their cultivation. This suggests that a particular type of wheat might be selected and grown specifically for the biofuel market, so as to secure the most favourable terms from the processors. As with any such specialisation, a crop grown in this manner would be less attractive to the general feed market in the event of it being rejected for use in biofuel.

The proximity of a farm to a processing plant is also a factor in that initial level of demand from each of these plants is likely to be limited and readily fulfilled from local growers who can avoid the additional cost of transporting a crop from further afield.

All types of sugar beet can be used for processing into ethanol. The first plant to be built for this purpose at Wissington in north west Norfolk is now being converted so as to produce biobutanol instead. Due to its weight and cost of transport, location has always been critical to sugar beet production, and this could be of significance in the biofuel market, as only one processing plant is being developed at the present time.

Both bioethanol and biobutanol can only be used as minor additives to petrol within standard engines, as the main problem is that a higher ethanol mix would damage gaskets and hoses. There is therefore no scope for small-scale or on-farm processing, as in the case of biodiesel, and growers will need to depend on the commercial market. This situation is could change in time, when engine modifications become available and, as in Brazil, higher grades of bioethanol will be used.

Biobutanol

It is also possible to use wheat and sugar beet to produce biobutanol as an alternative to bioethanol. It has the advantage that it may already be used in double the quantity when mixed in petrol, at up to 10%, within current regulations. The only locally grown feedstock from which this is currently due to be undertaken commercially is sugar beet.

Bio-oil

Bio-oil i is the liquid created during pyrolysis, a process operated in the more efficient heat and power plants, as mentioned in chapter 4. The oil is generally consumed within the plant itself, but it could be extracted and used to drive modified diesel engines.

Biomethane

Methane gas can be extracted from organic matter through anaerobic digestion. The most practical source of material for this is sewage and other wastes or manures. The potential for using this for commercial heating or power generation is mentioned in chapter 4, but also it is possible to operate individual systems and to harness the gas as a fuel for use in modified vehicles. The attractions of doing so are considerable:

- The gas can be made from a wide range of organic waste materials, notably livestock slurry and manure, the disposal of which is otherwise becoming increasingly regulated and difficult.
- If not harnessed in some way, those wastes would naturally give off a high volume of methane, which is one of the most powerful forms of greenhouse gases.
- The digestion process gives off a mix of methane and carbon dioxide ,which can be converted into a fuel that is far more efficient than bioethanol or biodiesel.
- The remaining residue can be used as an environmentally benign form of compost or fertiliser.
- Biomethane can be mixed with natural gas.

The process is being used extensively in a number of countries, particularly in the Far East and in Sweden, but has not as yet been developed in the UK. The main reasons for this lack of interest are:

- Unlike biodiesel and bioethanol, methane cannot be used in existing vehicle engines without major modifications.
- Vehicle manufacturers are unwilling to offer such modifications until there is evidence of a sufficient demand, while fuel distributors are disinclined to provide retail facilities for methane unless there were to be enough potential users.

● As a consequence, no commercial organisations are becoming engaged in working with biomethane, unlike other biofuels, which have attracted the interest of major processors.
● As the concept of biomethane covers a number of issues, including waste disposal, environmental enhancement and transport and heating fuels, it does not come under any specific Government department.

For the present, the use of methane is seriously limited by the problem of there being no ready source of modified car engines and no public facilities for refuelling. There is no indication that commercial processors in the UK will be taking in farm and other wastes for conversion into biomethane, in the way that they now do with feedstocks for the other biofuels. The technology and equipment for micro-processing methane does exist, and could provide an alternative solution to the growing problem of spreading of slurries and manure, but until the political and commercial situation changes, its use is likely to be limited to individual enthusiasts rather than provide a viable form of on-farm diversification.

Action points

● Check with corn merchants or direct with fuel processing companies as to the availability of contracts for supplying wheat, oilseed rape or sugar beet as an energy crop within the specific area.
● Assess the advantages or otherwise of committing all or part of future crops to the offered contracts.
● Check the availability of ECAP.
● Check the proposed terms for carbon certification and other environmental requirements.
● Consider what crop varieties are most beneficial.
● Assess the potential of using setaside land for all or part of the proposed production.
● Consider the potential for producing biodiesel and biogas on a farm scale, by analysing the costs of investment in acquiring equipment and time required in operating it and in fulfilling regulatory requirements. Also consider the disposal of by-products.
● Re-assess the market situation each year, in terms of the expansion of energy crop production within the UK, the outlook for sales of crops into conventional food markets and the threats from imported feedstocks.

Appendix 1

Trade organisations, government agencies and other sources of information

Biofuels

British Sugar
Processor
Sugar Way
Peterborough PE2 9AY
Tel: 01733 563171
www.britishsugar.co.uk

Greenergy International Ltd
Processor
198 High Holborn
London WC1V 7BD
Tel: 020 7404 7700
www.greenenergy.com

Green Fuels
Manufacturer
Unit 4, Chelworth Business Park
Crudwell
Malmesbury SN16 9SG
Tel: 01666 575002
www.greenfuels.co.uk

Green Spirit Fuels Ltd
Processor
Henstridge Trading Estate
Templecombe
Somerset BA8 0TN
Tel: 01963 365259
www.greenspiritfuels.com

Springdale Renewable Energy Ltd
Processor
Springdale Farm
Rudstone
Driffield YO25 4DJ
Tel: 01262 421106
www.springdale-group.com

Biomass

Balcas
Contractor
Laragh
Enniskillen BT94 2FQ
Tel: 028 6632 3003
www.balcas.com

Bidwells
Consultancy
5 Atholl Place
Perth PH1 5NE
Tel: 01738 630666
www.bidwells.co.uk

Biojoule
Pellet manufacturer
www.biojoule.co.uk

Biomass Industrial Crops Ltd
Contractor
Kingsmill Road
Cullompton EX15 1BJ
Tel: 01884 35899
www.bical.net

Coppice Resources Ltd
Contractor
LS8, Armstrong House
First Avenue
Doncaster Airport DN9 3GA
Tel: 01302 623220
www.coppiceresources.co.uk

Energy Crops Company
Contractor
32 Anyards Road
Cobham KT11 2LA
Tel: 01932 584455
www.energy-crops.com

Energy Power Resources Ltd
Power plant operator
Unit 6
Deben Mill Business Centre
Old Maltings Approach
Woodbridge IP12 1BL
Tel: 08450 51051
www.eprl.co.uk

ESD Biomass
Consultancy
Overmoor Farm
Neston SN13 9TZ
Tel: 01225 816866
www.esdbiomass.co.uk

Organic Power Ltd
Biogas equipment and consultancy
Gould's House
Horsington BA8 OEW
Tel: 01963 371300
www.organic-power.co.uk

Scottish Biofuels
Coppice contracting and processing
Castlebridge Business Park
Gartlove
Nr Alloa FK10 3PZ
Tel: 01259 733810
www.scottishbiofuel.co.uk

Renewable Energy Growers
Contractor
Unit 2
Market Weighton Business Centre
Becklands Park
York Road
Market Weighton
York YO43 3GL
Tel: 01430 871888
www.energycrop.co.uk

Renewable Fuels Ltd
Supply contractor
The Harrops
The Menagerie
Escrick
York YO19 6ET
Tel: 01904 720574
www.renewablefuels.co.uk

Rural Generation Ltd
Processor
65-66 Culmore Road
Londonderry BD48 8JE
Tel: 028 7153 8215
www.ruralgeneration.com

Springdale Renewable Energy Ltd
See: Biofuels above

Talbotts
Boiler manufacturer
Drummond Road
Astonfields Industrial Estate
Stafford ST16 3HJ
Tel: 01785 213366
www.talbotts.co.uk

TV Energy
Consultancy
Liberty House
New Greenham Park
Newbury RG19 6HS
Tel: 01635 817420
www.tvbioenergy-coppice.co.uk

Welsh Biofuels
Contractor
32 Chilcott Avenue
Brynmenyn Industrial Estate
Brynmenyn
Bridgend CF32 9RQ
Tel: 01656 729714
www.welsh-biofuels.co.uk

Wind power

Airtricity
Developer
29A Union Street
Greenock PA16 8DD
Tel: 01475 892 344
www.airtricity.com

Amec
Developer
65 Carter Lane
London EC4V 5HJ
Tel: 020 7634 0000
www.amec.com

British Wind Energy Association
Industry body
1 Aztec Row
Berners Road
London N1 0PW
Tel: 020 7689 1960
www.bwea.com

Country Guardian
Campaign group
www.countryguardian.net

D M Energy
Consultancy
23 The Gallops
Lewes
Sussex BN7 1LR
Tel: 01273 474446

Ecotricity
Developer
Axiom House
Station Road
Stroud
Gloucestershire GL5 3AP
Tel: 01453 769326
www.ecotricity.co.uk

ETSU
See: Future Energy Solutions

Fisher German
Consultancy
40 High Street
Market Harborough LE16 7NX
Tel: 01858 410 200
www.fishergerman.co.uk

Future Energy Solutions
Consultancy
AEA Energy Solutions
The Gemini Building
Fermi Avenue
Harwell International Business Centre
Didcot
Oxfordshire OX11 0QR
Tel: 0870 190 6374
www.future-energy-solutions.com

Merchant Wind Power
See: Ecotricity

National Wind Power
See: RWE npower

RWE npower
Developer
Trigonos
Windmill Hill Business Park
Whitehill Way
Swindon SN5 6PB
Tel: 01793 877777
www.rwe.com

Smiths Gore
Consultancy
12 Bernard Street
Edinburgh EH6 6PY
Tel: 0131 554 2211
www.smithsgore.com

Wayleaves.com
Consultancy
11 Earlswood Road
Cardiff CF14 5GH
Tel: 029 2076 6251
www.wayleaves.com

Appendix 2

Glossary

Anaerobic digestion
A process of controlled decomposition of animal and food wastes to produce methane gas.

Biobutanol
An industrial alcohol produced from crops such as sugar beet and usable as a mix in conventional petrol.

Bioethanol
Ethanol that has been produced by the biological fermentation of plant material, notably sugar, maize and wheat, and usable as a mix in conventional petrol.

Biodiesel
A diesel fuel for vehicles that can be produced from oilseed rape and blended with conventional diesel or used in its entirety.

Bio-Energy Capital Grants Scheme
Providing Government funding towards developing markets for biomass.

Bio-Energy Infrastructure Scheme
Grant aid for developing supply chains and infrastructure for converting biomass into heat and power.

Biofuel
Combustible fuels, including bioethanol and biodiesel (qv), produced by the biological fermentation of plant material.

Biogas
Methane gas that has been extracted from organic matter through anaerobic digestion.

Biomass
Crops and residues of timber and some agricultural by-products that can be used as fuel, usually when chipped, to produce heat and power or be processed into methane gas. Crops include coppice and miscanthus (qv) and residues tend to derive from timber felling and processing and arboricultural debris.

Biomethane
Specific description of biogas (qv).

Bio-oil
A combustible fuel derived during the process of pyrolysis (qv).

Carbon sequestration
The ability of trees to absorb carbon dioxide from the atmosphere while growing and to retain it until such time as their timber is burnt or decomposed.

CHP (Combined Heat and Power)
Process whereby water is heated to produce steam that is used to drive a turbine to generate electricity. The steam and the combustion gases from the fuel are then harnessed to heat cold water, which can be circulated as a source of space heating for local premises.

Climate Change Levy
A Government levy imposed upon commercial and institutional users of electricity other than that which is certified as being derived from a renewable source.

Climate Change Programme
A series of measures introduced by the European Union in 2000 with the aim of reducing national levels of carbon emissions.

Clusters
Term used to describe a group of wind turbines.

Co-firing
A policy introduced within the Renewables Obligation (qv) requiring power stations to mix a specified percentage of biomass material with coal.

Coppice
A timber crop raised from shoots produced from cut stumps of broadleaved species, notably hazel, chestnut and oak. The wood can be harvested in cycles of between 6 and 30 years according to variety and cut or chipped to produce woodfuel.

Energy Crops Aid Payment
An annual grant made through the EC to growers of crops that are processed into biomass and biofuels.

Energy Crop Scheme
Grant aid provided through Defra for the establishment of short rotation coppice and miscanthus (in Wales: Farm Enterprise Grant Scheme).

Gasification
A process when the burning of fuel is enhanced by passing air or steam through it to produce a mixture of carbon monoxide and hydrogen to drive a turbine.

kW
1000 Watts (qv)

Landfill gas
A mixture of methane and carbon dioxide produced through the decomposition of biodegradable wastes when stored underground, such as in filled municipal waste disposal sites.

Levy Exemption Certificates
Issued by Ofgem to suppliers of electricity using power from a renewable source.

Load factor
A measure of the percentage of time in which a wind turbine may be generating electricity.

Local grid
Network for distributing electricity within a locality, often when in a rural area by means of overhead lines, and generally operating at 11 or 33kV. Electricity generated from renewable resources is connected through a substation to the local grid as opposed to the national grid, which is used as a network between commercial suppliers.

Low Carbon Buildings Programme (LCBP)
Government instrument within the Communities Renewables Initiative (in Scotland: Scottish Communities and Householder Renewables Initiative; in Wales: Wood Energy Business Scheme) providing funding for implementing microgeneration schemes (qv).

LPG
Liquid Petroleum Gas. A transport fuel produced from "natural" gas, derived from fossil fuel sources such as the North Sea fields.

Microgeneration
System of generating power and / or heat for individual buildings or groups of buildings as opposed to linking the output to a local or national network.

Miscanthus
Also known as elephant grass, this bamboo-like plant can be harvested annually for use as a biomass fuel as well as other purposes such as equine bedding.

mW
1000kW (qv)

mWe
1000kW of electrical power

mWh
1000kW of thermal power, or heat. Also used to express the number of hours over which electrical power is being produced.

National grid
See Local grid

NETA
New Electricity Trading Arrangements introduced in 2001 with the purpose of evening out the supply and demand of electricity, and which applies to all generators other than those using renewable resources.

Non-Fossil Fuel Obligation
Government instrument replaced in 2002 by the Renewables Obligation (qv) as an initial means of implementing the Climate Change Programme.

Photovoltaics
Solar photovoltaics is a process still under commercial development whereby (sun)light can be converted directly into electricity by means of silicon cells.

PPS22
Planning Policy Statement 22 giving Government guidance to Local Planning Authorities on rural issues including specifically renewable energy.

Pyrolysis
A means of raising the efficiency of generating electricity from burning biomass and other fuels. The fuel is heated without air or steam so that it decomposes rather than burns and produces combustible gases, which are passed through a turbine and also probably a heat exchanger.

Renewables Obligation
A Government instrument introduced in 2000 effectively requiring suppliers of electricity to deliver a specified percentage of power accredited with Renewables Obligation Certificates (qv).

Renewables Transport Fuel Obligation
A Government instrument to be implemented in 2008 requiring that a specified percentage of biofuel be included in diesel and petrol and accredited with Renewables Obligation Certificates (qv).

ROCs
Renewables Obligation Certificates required by suppliers of electricity to demonstrate that a statutory proportion of that electricity has been generated from renewable resources or is otherwise supported by Certificates issued by Government to cover any prevailing shortfall.

RVOs
Recovered Vegetable Oils used as a source of fuel additive and comprising oils used in catering that have been recycled.

Setaside
An arrangement whereby farmers must exclude each year a specified percentage of their land from the cultivation of food crops as part of the conditions for receiving aid payments.

Short Rotation Coppice
Small dimension timber, such as willow, grown in densely planted rows specifically for periodic harvesting as a material for fuel.

Strategic Search Areas
Locations defined by the planning authorities as being appropriate for the potential development of wind clusters.

TAN 8
Acronym for Technical Advice Note 8 adopted by the Welsh Assembly and defining Strategic Search Areas (qv) for the siting of potential wind clusters.

Watt
Basic unit of electrical power when producing one volt in a current of one ampere.

Wet biomass
Material such as animal slurries and food waste that can be converted into a biogas by means of anaerobic digestion (qv).

Index

A

Anaerobic digestion 32, 41
Anemometer 20
ARBRE 29, 30

B

BABFO 39
Beaufort scale 14
Biobutanol 41
Biodiesel 40, 42
Bio-energy Capital Grants Scheme 27
Bioethanol 38, 40, 41, 42
Biofuel 3, 4, 9, 33, 38
Biomass 4, 9, 11, 25, 26, 27, 28, 33
Biomass Task Force 28
Biomethane 41, 42
Bio-oil 41

C

Civil Aviation Authority 19
Climate Change Levy (CCL) 3, 8, 23, 26, 28, 30
Climate Change Programme 2, 3, 5, 6, 23, 25, 28
Co-firing 11, 25, 26, 29, 30, 33
Combined Heat and Power (CHP) 11, 25, 26, 27, 28, 29, 30, 31, 35
Common Agricultural Policy (CAP) 2, 8, 9, 28, 32, 35
Coppice 25, 26, 27, 30, 34, 35, 39 (and see: Short Rotation Coppice)

D

Department for Food and Rural Affairs (Defra) 26
Department of Trade and Industry (DTI) 15, 21, 24, 29
Designated areas 8, 18, 22

E

Emissions Trading Scheme 6, 26
Energy Crops Aid Payment (ECAP) 27, 39, 40
Energy Crops Scheme (ECS) 26, 27, 30, 31, 34, 39
Energy Review 2, 3, 8, 15, 19, 23, 39
Energy Saving Programme 14, 20
Environmental Impact Assessment 22
Environmental Statement 22

F

Farm Enterprise Grant Scheme 26
Forestry 10, 33
Forestry Commission 10, 34

G

Gasification 30
Ground Source Heat 12

H

Hydroelectrics 3, 6, 12
Hydrogen 7

K

Kyoto Agreement 2, 6, 37

L

Landfill gas 33
Landscape Character Assessments 22
Levy Exemption Certificate 8
Liquified Petroleum Gas 37, 38
Load factor 15
Local Grid 13, 15, 17, 21, 24
Low Carbon Buildings Programme (LCBP) 23, 24

M

Ministry of Defence (MoD) 19
Miscanthus 25, 26, 28, 30, 31, 32, 35, 39

N

National Grid 8, 17
(and see: Local Grid)
New Electricity Trading Arrangements (NETA) 7
Nitrate Vulnerable Zones (NVZs) 32
NOABL 15, 19
Non-Fossil Fuels Obligation (NFFO) 3, 7
Northern Ireland Executive 23
NPPG 6 22
Nuclear power 1, 3

O

Ofgem 3, 8
Oilseeds 10, 37, 39, 40

P

Pellets 11, 26, 29, 30, 31, 34
Photovoltaics 12
Planning 13, 18, 20, 21, 22
PPS 22 19, 22
Poultry litter 9, 32, 36
Pyrolysis 30, 41

R

Recovered Vegetable Oils (RVOs) 37, 38
Regional Development Agencies (RDAs) 21, 23
Renewables Obligation (RO) 3, 7, 28, 29, 31
Renewables Obligation Certificates (ROCs) 7, 8, 11, 15, 24, 30, 34, 38
Renewable Transport Fuel Obligation (RTFO) 38, 39, 40
Residues 10, 33
Royal Institution of Chartered Surveyors (RICS) 23
Royal Society for the Protection of Birds (RSPB) 19, 21

S

Sawdust 11, 34
Scottish Executive 21, 23
Setaside 10, 27, 39, 41
Short Rotation Coppice (SRC) 4, 25, 26, 27, 28, 30, 31, 39
Solar panels 7, 12
SPP 6 22
Strategic Search Areas (SSAs) 21
Straw 9, 28, 31, 32
Sugar beet 38, 40, 41

T

TAN 8 21, 22
Timber 10, 11, 34, 35

W

Waste 11, 12, 25, 30, 31, 33, 34, 35, 41
Welsh Assembly 21, 23
Wet biomass 33
Wheat 35, 41
Willow 9, 27, 28, 29, 30, 31
Wind turbines 4, 6, 7, 8, 9, 10, 13, 14, 15, 16, 17, 18, 20
Woodchips 4, 11, 29, 30, 31, 34
Wood Energy Business Scheme 35
Woodfuel 11, 26, 30, 34